Brian G. Henington

INTRODUCTION TO WILDLAND FIREFIGHTING

KendallHunt
publishing company

Cover image by Brian Henington

Kendall Hunt
publishing company

www.kendallhunt.com
Send all inquiries to:
4050 Westmark Drive
Dubuque, IA 52004-1840

ISBN 978-1-4652-7405-2

Printed in the United States of America

About the Author

Brian Henington began his wildland firefighting career in 1994 as a seasonal firefighter with the New Mexico Forestry Division—Las Vegas District. He served three years as a seasonal firefighter before he was hired as a crew supervisor with the Forestry Division's Inmate Work Camp Program. Brian was promoted to Camp Supervisor in 1999 and continued in that role until 2002. After his service with the New Mexico Forestry Division, Brian was hired as the Deputy Director of Field Operations for the New Mexico State Land Office. In addition to his normal duties, Brian was tasked with developing a highly successful prescribed fire and emergency response program. Brian severed 8 years in this position until he was hired as a full time Fire Science Instructor with Central New Mexico Community College. Brian has recently returned to the New Mexico State Land Office as an appointed official. He serves as the Assistant Commissioner of Public Lands—Field Operations and Natural Resources Division. Commissioner Henington continues to teach in a part-time status and remains active in college instruction, emergency response activities, prescribed fire activities and wildland fire management.

Brian began teaching college courses at Albuquerque Technical and Vocational Institute (now Central New Mexico Community College) in 2000. He has instructed over 2,500 students in the areas of wildland fire suppression, incident command system, prescribed fire, fire ecology and leadership. Brian recently developed a new curriculum for the Fire Science Program—Wildland Fire Concentration, at Central New Mexico Community College.

During his more than 20 years of firefighting, Brian has or has held the qualification of: Incident Commander (Type 3, 4 & 5), Safety Officer, Division/Group Supervisor, Task Force Leader, Strike Team Leader, Information Officer, Crew Boss, Engine Boss, Firing Boss, Felling Boss, Faller C, Prescribed Fire Burn Boss and Ignition Specialist.

Brian attended college on a football scholarship and played 4 years as a quarterback and slot back at New Mexico Highlands University. He was awarded a bachelor or arts degree (double major) in Criminology and Political Science. After completing his bachelor's degree, Brian served as a graduate assistant at New Mexico Highlands University. Brian was awarded a Masters of Art Degree in Public Affairs. He later returned to receive the degree of Masters of Business Administration. Because Mr. Henington teaches in the fire science field, he felt it was important to maintain a degree in the field. In 2015, Professor Henington was awarded a Fire Science Degree—Wildland Concentration from Central New Mexico Community College.

Brian lives in Rio Rancho, New Mexico and is married to Ms. Shauna Henington. The couple has three children: Bailee, Traigh and Ashton. On his free time, Brian is an avid hunter, sportsman and football fan.

CHAPTER 1

OVERVIEW OF WILDLAND FIRE MANAGEMENT

Learning Outcomes

- Explain the key concepts used in wildland fire suppression.
- Describe the wildland firefighter qualification system.
- List three (3) wildland fire references used to assist firefighters.
- Identify the key wildland fire agencies or organizations and explain their role in wildland fire suppression.

Courtesy of Willie Lucero

Key Terms

Incident Response Pocket Guide
Position Task Book
Red Card
Wildfire
Wildfire Suppression
Wildland Fire
Wildland Fire Management

OVERVIEW

The management of wildland fires requires a large force of trained, equipped, prepared, and physically fit men and women. This organization involves thousands of firefighters across the United States. Professional wildland firefighters may hold career positions or combat fires only during the summer months. The wildland fire community is made up of federal, state, county, tribal, city, volunteer, and private individuals who must meet minimum national standards to participate in wildland fire suppression.

The national standards identifying the qualification, training, and physical fitness requirements are intended to establish a consistent certification process that should be used throughout the country. Wildland fire suppression is dangerous and dynamic. Firefighters who are properly trained and qualified will be able to safely and effectively perform their jobs, while reducing the potential for injury or major accidents, including line-of-duty fatalities.

KEY CONCEPTS

The following terms are key concepts used in wildland fire management:

- **Wildfire:** An unplanned, unwanted wildland fire including unauthorized human-caused fires, escaped wildland fire use events, escaped prescribed fire projects, and all other wildland fires where the objective is to put the fire out.[1]
- **Wildfire Suppression:** An appropriate management response to wildfire or prescribed fire that results in curtailment of fire spread and eliminates all identified threats from the particular fire.[2]
- **Wildland Fire:** Any non-structure fire that occurs in vegetation or natural fuels. Wildland fire includes prescribed fire and wildfire.[3]
- **Wildland Fire Management:** The activities taken by wildland fire managers to suppress, manage, and/or control wildland fires.
  - » This concept also includes the actions taken by fire managers to direct and guide individuals participating in fire suppression activities.
  - » The activities taken to reduce the potential for catastrophic fire (e.g., fuel mitigation, thinning, prescribed fire, etc.) or activities to improve forest health and ecosystem management.

WILDLAND FIREFIGHTER QUALIFICATIONS

Wildland fire agencies and most structural fire departments follow and adhere to national wildland fire qualifications and standards. These qualifications are identified and determined by the **National Wildfire Coordinating Group** or NWCG. The NWCG provides the training, experience and physical fitness standards for each position used on a wildland fire. The purpose of NWCG is identified below:

---

[1] National Wildfire Coordinating Group, *Glossary of Wildland Fire Terminology* (Boise: National Wildfire Coordinating Group, 2014), 85.

[2] Ibid.

[3] Ibid., 186.

NWCG provides national leadership to develop, maintain, and communicate interagency standards, guidelines, qualifications, training, and other capabilities that enable interoperable operations among federal and non-federal entities. NWCG will facilitate implementation of approved standards, guidelines, qualifications, and training.[4]

Qualifications and Experience Documentation

Wildland firefighters use an effective and useful tool to document work tasks, competencies, and experience. This tool is referred to as a ***Position Task Book***. The *Position Task Book* is published by the NWCG for job positions used on wildland fires. The NWCG describes the *Position Task Book* as:

A document listing the performance requirements (competencies and behaviors) for a position in a format that allows for the evaluation of individual (trainee) performance to determine if an individual is qualified in the position.[5]

The task books are essential to document successful function and performance by the firefighter for each task identified in the task book. As a wildland firefighter, you should become familiar with the tasks identified in your task book and work closely with your supervisor to identify opportunities to successful perform these tasks. Note: Task books must be initiated by an authorized agency representative and must be evaluated by a qualified firefighter.

The illustration below is an example of a Position Task Book.

A Publication of the National Wildfire Coordinating Group

NWCG Task Book for the Positions of:

FIREFIGHTER TYPE 1 (FFT1)

INCIDENT COMMANDER TYPE 5 (ICT5)

(POSITION PERFORMANCE REQUIRED ON A WILDFIRE ASSIGNMENT)

PMS 311-14 JUNE 2009

Task Book Assigned To:

Trainee's Name: ______

Home Unit/Agency: ______

Home Unit Phone Number: ______

Task Book Initiated By:

Official's Name: ______

Home Unit Title: ______

Home Unit/Agency: ______

Home Unit Phone Number: ______

Home Unit Address: ______

Date Initiated: ______

The material contained in this book accurately defines the performance expected of the position for which it was developed. This task book is approved for use as a position qualification document in accordance with the instructions contained herein.

Figure 1-2

Common Tasks for FFT1 and ICT5

This task book contains the tasks for both Firefighter I (FFT1) and Incident Commander Type 5 (ICT5). The common tasks for both positions are listed first. The tasks specific to Incident Commander Type 5 are listed following the common tasks. The FFT1 tasks only need to be completed once. The FFT1 and ICT5 PTBs can be initiated at the same time; the tasks can be completed simultaneously. A Verification/Certification page is included in this PTB for each position.

Common Tasks for FFT1 & ICT5 pages 7 – 16 (Tasks 1 – 39)
ICT5 Specific Tasks pages 17 – 22 (Tasks 40 – 60)

Competency: Assume position responsibilities.
Description: Successfully assume role of Firefighter and/or Incident Commander and initiate position activities at the appropriate time according to the following behaviors.

TASK	CODE	EVAL. RECORD #	EVALUATOR: Initial & date upon completion of task
Behavior: Ensure readiness for assignment.			
1. Obtain complete information from dispatch upon assignment. • *Incident name* • *Incident order number* • *Request number* • *Incident phone number* • *Reporting time* • *Reporting location (drop point)* • *Transportation arrangements/travel routes* • *Contact procedures during travel (telephone/radio)*	O		
2. Bring adequate personal gear and effects within established weight requirements.	O		
3. Follow safety procedures for foot travel and for transporting personnel and equipment (loading, riding, and unloading). • *Vehicles* • *Boats* • *Helicopters* • *Large transport aircraft* • *Small fixed-wing aircraft*	O		

Evaluate the numbered tasks ONLY. DO NOT evaluate bullets; they are provided as examples/additional clarification.

7

Figure 1-3

[4] National Wildfire Coordinating Group, *Charter* (Boise: National Wildfire Coordinating Group, 2013), 2.

[5] National Wildfire Coordinating Group, *Glossary*, 138.

Physical Fitness

Wildland firefighters utilize a standard assessment test to evaluate the fitness level of every firefighter. The test is called the **Pack Test** or **Work Capacity Test**. Each job position used on a wildland fire will require a specific physical fitness standard. As an entry-level firefighter, you will have to pass the test at an arduous level. The arduous level requires a three-mile walk with a 45 lb. pack/vest. This must be completed under 45 minutes. (Time may fluctuate based on the elevation where the test was administered.) The pack test is explained in detail in *Chapter 6: Firefighter Readiness and Core Essentials.*

Incident Qualification Card—Red Card

The **Incident Qualification Card (Red Card)** is a national qualification system used by members of the NWCG. The Red Card defines an individual firefighter's qualified position(s), as it relates to *National Incident Management System: Wildland Fire Qualification System Guide—PMS 310-1.* The Red Card also identifies trainee or target positons. The Red Card is produced by your host agency on a yearly basis, after you have successfully passed the work capacity test (pack test). Firefighters will be required to show their red card before engaging fires on federal and most state fires.

Firefighter Type 2

Entry-level wildland firefighters will function under the job position of Firefighter Type 2 (FFT2). This is the basic qualification used in the active suppression of wildland fires. The intent of this book is to prepare you for the Firefighter Type 2 position. Completion of the book and required NWCG courses for FFT2 will not automatically certify you for the position. You must *successfully* perform the duties and competencies of the positon on a fire. It may take an entire fire season or more to become fully certified as a Firefighter Type 2.

Firefighter Type 2 training requirements (as established by NWCG).

- I-100 Introduction to ICS
- L-180 Human Factors on the Fireline
- S-190 Introduction to Wildland Fire Behavior
- S-130 Firefighter Training
- IS-700: NIMS: An Introduction
- RT-130 Annual Fireline Safety Refresher (done once a year after you have completed all of the above training.)

Firefighter Type 2 Supervision

The direct supervision of the Firefighter Type 2 usually falls under the direction of a qualified Firefighter Type 1–FFT1. Fire agencies may utilize other supervision structures, but this is the most common used system throughout the country. If you are hired to work on a 20-person hand crew, your supervisor will be a Firefighter Type 1 (Squad Boss). If you work on a fire engine, your immediate supervisor will also be a qualified Firefighter Type 1 (often called an Engine Operator) or you may report directly to the individual responsible for the fire engine—an Engine Boss.

WILDLAND FIRE REFERENCES

Wildland fire management has hundreds of concepts, terms, and standards that firefighters must understand and use when functioning on a wildland fire. It is difficult for experienced firefighters to remember every concept, rule, or guideline. To help improve the comprehension of these essential concepts, fire managers have developed several fireline references and resources. As an entry-level firefighter, you should become extremely familiar with the reference material identified below.

Incident Response Pocket Guide (IRPG)

The most important fireline reference material used by wildland firefighters, ***The Incident Response Pocket Guide*** (IRPG) was designed to provide firefighters with best practices, safety protocols, hazard identification and mitigation, and common concepts. The IRPG is designed to fit in the pocket of your fire shirt. You should carry the IRPG at all times and become familiar with it. In addition, the IRPG is a very useful training tool used by fire supervisors to train entry-level firefighters and refresh concepts for seasoned firefighters.

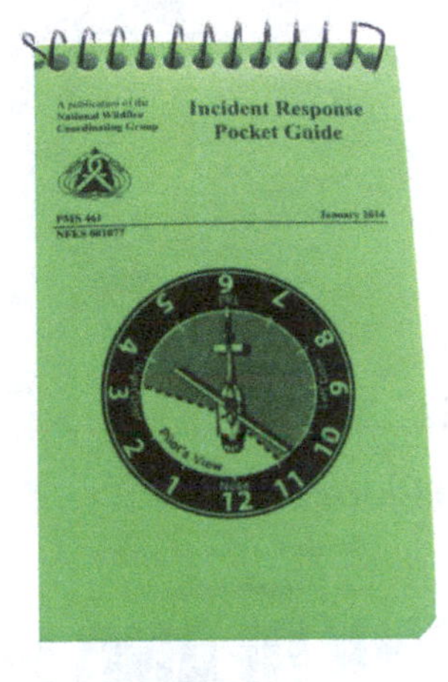

Courtesy of Brian Henington

Figure 1-4

The IRPG is periodically updated. Ensure you have the latest version.

IRPG Features:

- Core Firefighting Responsibilities
- Operational Engagement
- Specific Hazards
- Aviation
- Other References
- Emergency Medical Care
- Rules of Engagement

Wildland Fire Incident Field Management Field Guide

The *Wildland Fire Incident Management Field Guide* is a new reference material to the wildland fire arena. This reference material has replaced the very popular and useful reference tool referred to as the *Fireline Handbook.*

The *Wildland Fire Incident Management Field Guide* provides incident management requirements and responsibilities as well as providing detailed information related to fire suppression concepts and safety measures/practices.

This reference tool can be purchased as a hard copy or can be downloaded for free.

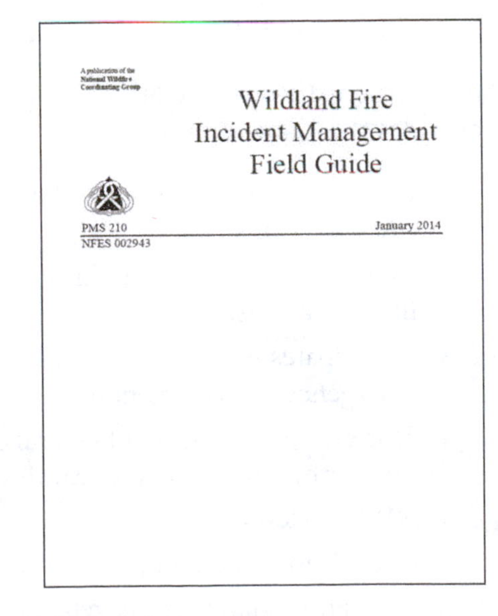

Figure 1-5

National Incident Management System: Wildland Fire Qualification System Guide—PMS 310-1

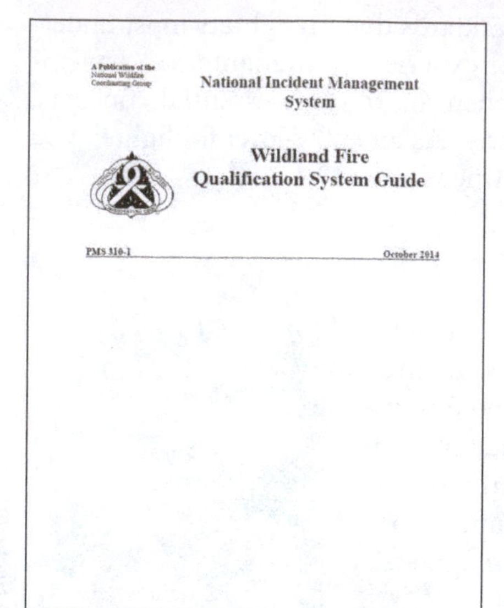
A Publication of the National Wildfire Coordinating Group

National Incident Management System

Wildland Fire Qualification System Guide

PMS 310-1 October 2014

Figure 1-6

The National Incident Management System: Wildland Fire Qualification System Guide—PMS 310-1 is published and maintained by the NWCG to provide a consistent and precise qualification standard for job positions used on wildland fires.

The manual is specifically designed to provide a system for national response. If a firefighter is dispatched from North Carolina to support fire operations in Idaho, that firefighter must meet the standards identified in this standards guide. It is expected that the firefighter from North Carolina will meet the same requirements and qualifications as a firefighter from Idaho. In addition, this standard guide also provides qualification standards for prescribed fire activities.

This manual provides the qualification standards for over 100 positions used on wildland fires. It identifies the required training, experience, physical fitness level, currency requirements, and recommended supporting training.

WILDLAND FIRE AGENCIES AND ORGANIZATIONS

As mentioned before, the wildland fire community is made up of thousands of firefighters and hundreds of fire agencies located throughout the nation. This section provides a brief overview of the primary agencies involved in suppression and support activities.

National Wildfire Coordinating Group (NWCG): www.nwcg.gov

As mentioned above, the NWCG is one of the most important wildland fire organizations. The NWCG provides the following functions:

- Develops and proposes standards, guidelines, training, and certification for interagency wildland fire operations.
- Maintains approved standards, guidelines, training, and certification for interagency wildland fire operations.
- Participates in the development of operational standards and procedures for non-fire incident and emergency management to ensure consistency and interoperability.[6]
- This group is made up of several organizations. These organizations have input and representation for qualification and training standards. The NWCG is located in Boise, Idaho.
- NWCG Membership
  - » US Department of Agriculture—Forest Service
    - – Fire Aviation Management and
    - – Wildland Fire Management Research, Development, and Application

[6] National Wildfire Coordinating Group, *Chapter*, 2.

» US Department of Interior
  - Bureau of Land Management
  - Fish and Wildlife Service
  - National Park Service
  - Bureau of Indian Affairs
» US Department of Homeland Security
  - Federal Emergency Management Agency—US Fire Administration
» National Association of State Foresters
» Intertribal Timber Council
» International Association of Fire Chiefs

National Interagency Fire Center (NIFC): www.nifc.gov

The National Interagency Fire Center (NIFC) is considered the national support center for the wildland fire community. NIFC is also located in Boise, Idaho. The purpose of this entity is to provide coordination among wildland fire agencies and reduce the duplication of specific efforts and activities. NIFC is essential in providing intelligence, information, safety standards, and policies.

Federal Agencies

US Department of Agriculture—Forest Service (USFS): www.fs.fed.us

The US Forest Service is the largest employer of wildland firefighters across the nation and maintains a very large array of firefighting resources and equipment. The USFS manages 154 national forests and 20 national grasslands in addition to their fire suppression/support responsibilities.

US Department of Interior—Bureau of Land Management (BLM): www.blm.gov

The Bureau of Land Management is responsible for the management of specific lands across the western United States. Encompassing more than 245 million acres, the BLM provides a wide range of management activities for its jurisdictional authority. In addition to its land management efforts, the BLM employees and maintains a large force of wildland firefighters, equipment, and resources to support fire operations throughout the country.

US Department of Interior—Fish and Wildfire Service (USFWS): www.fws.gov/fire

The US Fish and Wildlife Service—Fire Management Program is responsible for managing 150 million acres of jurisdictional authority in 560 national wildlife refuges and 38 wetland management areas. The agency is highly involved in fire management activities in their management area and also supports fire operations across the nation.

US Department of Interior—National Park Service (NPS): www.nps.gov/fire/wildland-fire

The National Park Service is responsible for managing and protecting its properties across the nation. It manages over 400 parks that encompass more than 80 million acres. In addition to its primary function, the NPS provides a highly functional wildland fire organization.

US Department of Interior—Bureau of Indian Affairs (BIA): www.bia.gov

The Bureau of Indian Affairs is responsible for the management and administration of 55 million acres located in federally recognized American Indian tribes and Alaska Natives areas. The BIA Division of Forestry and Wildland Fire Management is actively involved in fire management activities on property managed and owned by the department and supports fire activities across the nation.

National Weather Service (NWS): www.weather.gov

The National Weather Service is essential to wildland fire suppression activities. The fire weather component of the National Weather Service provides critical fire weather information, predictions, forecasts, and outlooks. It is one of the most important support agencies used by firefighters across the nation.

US Federal Emergency Management Agency (FEMA): www.fema.gov

FEMA's US Fire Administration is a contributing member of the National Interagency Fire Center and the National Wildfire Coordinating Group. It is primarily responsible for the development and delivery of Incident Command System and National Incident Management System training curriculum. FEMA will also support large-scale fire incidents and events.

Non-Federal Entities

National Association of State Foresters: State Forestry or Department of Natural Resources: The National Association of State Foresters is composed of 59 member agencies. The agencies involved include states and US Territories. State entities have jurisdictional authority for the lands they own or manage. Some state entities have a large force of wildland firefighters while others do not. The state agencies are highly effective and typically function under state laws or statutory authority.

- Alabama
- Alaska
- Arizona
- Arkansas
- California
- Colorado
- Connecticut
- Delaware
- District of Columbia
- Florida
- Georgia
- Hawaii
- Idaho
- Illinois
- Indiana
- Iowa
- Kansas
- Kentucky
- Louisiana
- Maine
- Maryland
- Massachusetts
- Michigan
- Minnesota
- Mississippi
- Missouri
- Montana
- Nebraska
- Nevada
- New Hampshire
- New Jersey
- New Mexico
- New York
- North Carolina
- North Dakota
- Ohio
- Oklahoma
- Oregon
- Pennsylvania
- Rhode Island
- South Carolina
- South Dakota
- Tennessee
- Texas
- Utah
- Vermont
- Virginia
- Washington
- West Virginia
- Wisconsin
- Wyoming
- American Samoa
- Virgin Islands
- Guam
- Puerto Rico
- Palau
- Fed. States of Micronesia
- Rep of the Marshall Islands
- Northern Mariana Islands

County Fire Departments: Some county fire departments across the country have an elaborate wildland fire division or program. County fire departments typically operate within their assigned counties; however, they can support national fire activities if they follow the standards identified in *NWCG 310-1*.

City or Municipality Fire Departments: Many cities or municipalities in the country have jurisdictional response areas that include wildland areas. These fire departments have the primary authority to suppress wildland fires. As in the case of county fire departments, city or municipal fire departments offer great career opportunities, benefits, and retirement.

Volunteer Fire Departments: A majority of wildland fires are handled at the volunteer level and never require assistance from state or federal resources. Volunteer fire organizations offer a great opportunity to support the safety of rural communities across the United States.

Private Companies: Wildland fire suppression is big business. Many private companies have been created to deal with certain aspects of wildland fire activities. These organizations can provide aircraft, operational resources, support functions, food catering, etc. Companies participating in suppression activities on federal and most state jurisdictional areas must be red-carded and meet the standards identified in *NWCG 310-1*.

Support Organizations

Military: Certain branches of the US military own and manage properties that are susceptible to wildland fire ignitions. These groups may have their own firefighters or work in conjunction and coordination with professional wildland fire organizations to manage fires. In addition, during extreme fire conditions, the military may be activated to support fire operations across the nation. The most common military organizations used to support fire operations are: Air National Guard, National Guard, Air Force Reserve, Army, Marines, and the Air Force.

Colleges/Universities: Colleges and/or universities often provide support functions to wildland fire agencies. The main intent of these educational institutions is to provide college degrees or technical training to aspiring firefighters. In addition, many fire agencies are now requiring a college degree to receive a promotion or full-time position. Finally, research universities and colleges can also be very useful to wildland fire agencies by providing research and data related to wildland fire, forestry, watershed health, ecosytem management, etc.

Wildland Fire Lessons Learned Center: www.wildfirelessons.net

The Wildland Fire Lessons Learned Center was established in 2002 and is located at the National Advanced Fire and Resource Institute (NAFRI) in Tucson, Arizona. This organization provides resources and reference material through web-based platforms. The main focus of this group is to provide tools and tactics to the wildland fire community to ensure safety and effectiveness in all field operations.

- *Six Minutes for Safety* is a safety program that provides daily briefings related to fire operations and safety concepts. Many wildland fire organizations use *Six Minutes for Safety* to conduct their daily safety briefings.
- *2 More Chains* is a quarterly publication that focuses on sharing information among wildland firefighters. The publication does a fantastic job of addressing safety concerns while developing and enhancing the firefighter's situational awareness.

Wildland Fire Leadership Development Program: www.fireleadership.gov

This program was created to address the area of leadership on wildland fires. The concept of leadership is an essential function to ensure effective operational activities and safe firefighting actions. The goal of the program is to drastically increase the number of competent and qualified fireline leaders. This organization manages, develops, and maintains the training curriculum for the "L" series of classes used by the NWCG.

- The Leadership Program identifies three core values. They are duty, respect, and integrity.
- "L" Series Training Includes:
  - » L-180 *Human Factors*
  - » L-280 *Followership to Leadership*
  - » L-380 *Fireline Leadership*
  - » L-381 *Incident Leadership*
  - » L-480 *Organizational Leadership*
  - » L-481 *IMT Leadership*
  - » L-580 *Leadership is Action*

InciWeb: www.inciweb.nwcg.gov

InciWeb is hosted by NWCG and provides information (by state) on active large wildland fires, prescribed fires and/or other emergency activities that wildland fire organizations are involved in.

SAFENET—Wildland Fire Operations Safety Reporting System: http://safenet.nifc.gov

SAFENET is a process used by wildland firefighters to identify activities that are unsafe and/or have jeopardized firefighter safety. The main focus of this program is to provide timely resolution to safety/health concerns.

Wildland Firefighter Foundation: www.wffoundation.org/

The Wildland Firefighter Foundation is a nonprofit organization that supports firefighters who have been injured in the line of duty or the families of fallen firefighters. The organization is located in Boise, Idaho and has proven to be essential in helping individuals dealing with devastating tragedies. The Wildland Firefighter Foundation also maintains the Wildland Firefighter Monument. Every wildland firefighter should make an effort to visit this wonderful monument and support its mission through donations.

In addition to other great programs, the Wildland Firefighter Foundation's has two main priorities:

- Help families of firefighters killed in the line-of-duty and to assist injured firefighters and their families.
- Honor and acknowledge past, present, and future members of the wildland firefighting community, and partner with private and interagency organizations to bring recognition to wildland firefighters.

SUMMARY

Wildland fire management is complex, dynamic, and involves thousands of firefighters, resources, organizations, and agencies. All of these groups work together to safely combat wildland fires. A standardized system is needed to ensure all individuals participating in wildland fire management are certified, experienced, and meet established physical fitness standards. Without these standards, injuries, accidents, or fatalities to firefighters would be greatly increased and firefighting efforts would not be organized or managed in a manner that is safe and effective.

KNOWLEDGE ASSESSMENT

1. Explain the difference between a wildfire and a wildland fire.
2. Define wildland fire management.
3. What tool do wildland firefighters use to document work activities, competencies, and experience?
4. What is the title of an entry-level firefighter?
5. Identify three references used to support wildland fire activities.
6. Out of the three reference materials, which one is considered the most important?
7. What wildland fire organization develops and maintains training standards and curriculum?
8. What is the primary function of the National Interagency Fire Center?
9. What role does the National Weather Service play in wildland fire management?
10. What are the three core values of the Wildland Fire Leadership Program?
11. What is *Six Minutes for Safety*?

EXERCISES

1. Research the state agency that is responsible for wildland fire suppression in your state. How is the agency established (e.g., state forestry or natural resource department)? How large is it? Provide a web link of the agency for your instructor.
2. Research the Wildland Firefighter Foundation using the Internet. What function of the organization really stood out to you?

BIBLIOGRAPHY

National Wildfire Coordinating Group. *Charter.* Boise: National Wildfire Coordinating Group, 2013.

National Wildfire Coordinating Group. *Firefighter Training, S-130.* Boise: National Wildfire Coordinating Group, 2003.

National Wildfire Coordinating Group. *Glossary of Wildland Fire Terminology.* Boise: National Wildfire Coordinating Group, 2014.

CHAPTER 2

THE MANAGEMENT OF WILDLAND FIRES

Learning Outcomes

- Define the management structure used on all emergency incidents throughout the United States and explain its use and significance.
- Describe the importance of the chain of command and how it is utilized on wildland fires.
- List the five (5) incident types and identify when they would be best utilized.
- Explain the five (5) major management activities (positions) around which the ICS is organized.
- Identify the three (3) management positions in the command staff and explain their roles on an emergency.

Courtesy of Brian Henington

Key Terms

Camp
Chain Of Command
Deputy
Divisions
Group
Helibase
Helispot
Incident
Incident Action Plans
Incident Base
Incident Commander
Incident Command Post (ICP)
Incident Command System
Incident Complexity
Incident Types
Overhead
Shifts
Span Of Control
Single Resources
Spike Camp
Staging Areas

OVERVIEW

The management of emergencies has become more complicated and complex in the last couple of decades. The amount of emergency personnel and resources needed to deal with some catastrophes can be massive, requiring hundreds if not thousands of responders. Disasters like Hurricane Katrina and Superstorm Sandy demanded a massive amount of emergency responders and support personnel from all over the United States to assist with the recovery activities. All responders must have a standardized management system to help coordinate and direct resources in a consistent and efficient manner.

Wildland fire agencies have been effectively using a standardized management tool to mobilize and respond to these emergencies since the 1970s. Large wildfires present unique challenges. They burn across thousands of acres, threaten homes, and require a large force of firefighters to manage and suppress them. To complicate the matter, more people are building homes in wild areas, which results in increased chaos when a fire is introduced into the picture. Fire managers must also support the needs of the firefighters and emergency responders. They supply food, water, showers, communications, vehicle repairs, and other necessities. In order to accomplish all of these goals, fire managers have instituted the use and application of a on scene management system called the **Incident Command System** (ICS).

The terrorist attacks of September 11th proved that not all emergency response agencies were able to deal with large scale and/or long duration events. Agencies were managing emergencies under their own management standards. One agency would operate in a manner confusing or unfamiliar to responders from assisting agencies. Responders were not able to communicate and coordinate activities in the most efficient and safe manner. Furthermore, Hurricane Katrina illustrated the lack of coordination and cooperation between agencies. There have been numerous emergency events throughout history that have demanded the need for emergency agencies to communicate, coordinate, and manage incidents in a system that all understand. Today, all emergency response agencies across the United States must use a standardized management system called the National Incident Management System, with the Incident Command System being the on-scene (boots on the ground) management tool.

After September 11th, agencies were mandated by a presidential declaration (*HSPD-5*) to adopt and coordinate all incident activities under the National Incident Management System and the Incident Command System (ICS). To participate in wildland fires, all firefighters must complete an online training through the Federal Emergency Management Agency (*I-100.B. Introduction to Incident Command System, ICS-100*) In addition, firefighters must also complete a self-paced class on the *National Incident Management System, An Introduction: IS-700.A*. The purpose of this chapter is not to focus or rehash the concepts in these two classes. The intent is to provide an introduction and focus on wildland fire so you can understand how the system works and the role you play in this system.

PERSONAL PERSPECTIVE

I was tasked several years ago with presenting an overview of the functional use of ICS and how it is implemented on wildland fires. The presentation was to several fire chiefs and fire managers. After the presentation, a deputy fire chief from a large city fire department started to criticize ICS. He made accusations that the system was ineffective and there were better systems available to deal with emergencies. I asked what he based his declaration on. He claimed that the fires that struck California in 2003 illustrated that ICS was not effective because it did not provide adequate tactical solutions. I looked around the room in shock. I wondered if he had not heard the introduction to the presentation! ICS is a management tool! It is not designed to teach the use of chainsaws or how to construct firelines. It is used to set objectives and we must use the appropriate tactics to meet the objectives. As wildland firefighters, we learn how to use tactics in suppression-related classes, not in incident-management classes. This example illustrated how many old school firefighters felt about the Incident Command System. In your career, you will be fortunate to witness ICS in its true form. You should not have people fighting the system because all responders must use it.

Another example of the inappropriate use of the ICS occurred on a large Bosque fire in central New Mexico. The fire started on the east side of the river and with high winds influencing activity, the fire jumped the river to the west side. There was active fire on both sides of the river. Many values were being threatened by the fire: homes, pipelines, transmission lines, etc. When our resources arrived, the scene was total chaos. The east side of the river was being managed by one incident commander. The west side of the river was being managed as a total different incident with a different incident commander. Resources were operating on different radio frequencies. They were not communicating effectively and most importantly, safety measures were not in place. We made the decision not to engage the fire until an adequate management structure was in place. This upset both incident commanders. We had 45 firefighters ready and willing to work, but we were not going to do so if we could not communicate to other resources or understand the intent of the commanders. Our responsibility as emergency responders is to bring order to the chaos. The illustration here is that we were not going to add to the chaos or confusion of this incident and jeopardize firefighter safety.

The issue above was finally rectified after several hours. We were tasked with taking the fire over and we immediately implemented ICS: one incident commander with established common frequencies and safety measures. Instead of separating the same incident into two different incidents, we used the ICS and its capability by establishing an operations section with two divisions—one for the east side of the river and one for the west side of the river.

THE SYSTEM

The Incident Command System (ICS) is an "all risk" on-scene management system. Not only is it used on wildland fires, it is used for other emergency incidents including pre-planned activities. The system does not provide tactical solutions to an incident. The system is a management tool. An incident is explained as "an occurrence either human-caused or natural phenomenon, that requires action or support by emergency service personnel to prevent or minimize loss of life or damage to property and/or natural resources."[1] Examples of emergencies that use ICS are identified below.

[1] National Wildfire Coordinating Group, *Glossary of Wildland Fire Terminology* (Boise: National Wildfire Coordinating Group, 2014), 104.

Courtesy of Brian Henington

Wildland Fires.

© Knumina Studios/Shutterstock.com

Structure Fires.

Courtesy of Brian Henington

Search and Rescue.

© Ben Carlson/Shutterstock.com

Hazardous Material Incidents.

© Martin Haas/Shutterstock.com

Tornado Recovery.

© Glynnis Jones/Shutterstock.com

Hurricane and Floods.

© Ken Tannenbaum/Shutterstock.com

Multi Casualty Events.

© Png Studio Photography/Shutterstock.com

Planned Events.

© Aspen Photo/Shutterstock.com

Parades/Concerts.

All emergencies present some type of confusion or chaos. The goal of the ICS and the person in charge is to bring control to a dangerous situation. The ICS was designed to help facilitate this concept by providing a clear, defined structure that allows us to understand what our role is, how we communicate and cooperate, and how we support the firefighters risking their lives battling these fires.

Below is a list of key components of the ICS:

1. ICS is a management tool.
2. ICS is flexible and adapts with the complexity of the incident. It grows as the incident elevates, and it shrinks and the incident decreases.
3. ICS utilizes and enforces the concept of chain of command and span of control.
4. Complexity is based on several factors, with the top priority being life safety.

5. ICS is an all-risk management system. It is used for all emergency activities; not just large or complex incidents. Responders will use the same system for a small, 1 acre wildland fire and as for the recovery efforts of a tornado. It is designed for all incidents.
6. The amount of personnel and resources needed is deemed by the complexity of the incident.

Incident Complexity

To understand incident type, we must first define what complexity is as it relates to emergency activities. **Incident complexity** defines the level of effort, coordination, cooperation, and management structure needed to effectively deal with a specific incident. Several factors many contribute to the complexity. On a wildland fire, complexity can increase or be elevated by several factors or considerations. They can involve some of the following (not all identified here and can be incident specific):

- Numerous air resources assigned to the same incident.
- Need to evacuate the public.
- Critical natural resource areas or concerns.
- Major infrastructure or facilities threatened.
- Difficult logistical needs.
- Extreme fire behavior.
- Size and location of the fire.

Five Core Activities of ICS

ICS is based on five core activities or responsibilities. The five core activities will be explained in detail further in the chapter. The five core activities include: (The quick definition is provided for you to remember in laymen's terms what each activity is responsible for.)

1. **Command:** responsible for the command and control of the incident.
2. **Operations:** "The Doers." Operations include the men and women who are responsible for the tactical activities on the emergency.
3. **Planning:** Evaluates and plans for incident needs. Plan for two hours from now, one day from now and/or two weeks from now.
4. **Logistics:** Supply you with what you need.
5. **Finance/Administration:** Those who pay you. Responsible for financial and administration requirements;

Organizational Terminology

The use of common terms in ICS is an important function of this system. We have to be consistent with the terms we use to ensure we facilitate any communication issues. ICS has a lot of terms that may be confusing to the entry-level firefighter. Over time, you will begin to understand and effectively use the appropriate term during the correct situation. An example of this: Most structure fire agencies use the word "tanker" to identify a piece of equipment that provides water to the incident. In wildland fire, the term "tanker" identifies aircraft (either an air tanker or helitanker) that delivers retardant or water to the fire through the air. If you want a water tender and order a tanker, you may receive an air tanker. You have to be clear and use the correct terminology to ensure your needs are effectively communicated.

Important Terms:

- **Deputy:** It's not uncommon for large fire organizations to have a Deputy Incident Commander. In order for this to occur, deputies must be as qualified as the person they work for. Other positions that may have deputies include Section Chiefs and or Branch Directors.
- **Shifts:** The amount of hours worked in a day; should adhere to the two hours worked to one hour of rest ratio. Shifts cannot exceed 24 hours, with the most common shifts being 12–16 hours an operational period.
- **Overhead:** Emergency responders in management positions. Reasonable for managing and directing certain activities. They can be considered mid-level management to top-level management.

Common Responsibilities

It is every firefighter or emergency responder's obligation to ensure the following common responsibilities occur prior to an assignment, once you arrive at the assignment, during the assignment, and when you leave. They include:

- Receive an assignment with the resource order number and directions to your incident.
- Ensure you bring specialized equipment. What job are you being dispatched to perform? For example, if you are dispatched as a chainsaw operator, then you need to bring your chainsaw (as long as you're not flying on a commercial jet) and the supporting chainsaw tools and equipment.
- Officially check in. Check-in is a formal process and should be done when you arrive at the incident. The process includes providing a manifest, resource number, and verification of your Incident Qualification Card (Red Card).

  You should check in at Incident Command Post or at any of these locations:
  - Staging areas
  - Base or camp
  - Helibase
  - Division or Group Supervisor for direct assignment.
- Use clear text on the radio. No 10 codes! 10–99 in the State of Wyoming refers to a Wanted or Stolen vehicle. 10–99 in the state of New Mexico means an officer is being held hostage. You can see how communication issues will occur if 10 codes are used. Talk on the radio as you would talk face-to-face.
- Ensure you obtain a briefing from your direct supervisor and if you are a supervisor, ensure you brief the people working for you.
- Ensure you are organized and have the appropriate tool(s) for the specific task.
- Work with the adjoining resources and brief your replacement at the end of your operational period.
- Complete all required forms and submit them on a regular basis.
- Demobilize according to the plan identified for you or your crew.

Chain of Command

The chain of command identifies the supervision structure and level of authority used in the ICS. The National Wildfire Coordinating Group identifies the chain of command as, "… the line of authority through which decisions are made, recommendations offered, and work assignments are given."[2] It is important for us to have one boss (this is called unity of command) to facilitate job assignments, safety considerations, and the appropriate communications. As an entry-level firefighter, you should clearly understand who your boss is and who his/her boss is. Run all decisions and concerns through your chain of command. (This may not always be an option, but is a good rule of thumb.)

Courtesy of Brian Henington

SPAN OF CONTROL

Three to seven resources with an ideal ratio of one supervisor managing five reporting elements.

Span of Control

The span of control is an essential element of the ICS. Fire managers have to ensure that each firefighter has adequate supervision. This allows for effective communication and direction. The ICS specifies span of control may vary from three to seven resources with an ideal ratio of one supervisor to five reporting elements. Some specialized and highly trained hand crews may have a larger span of control because of their capabilities and ability to effectively communicate. Figure 2-1 illustrates the span of control for a strike team of fire engines.

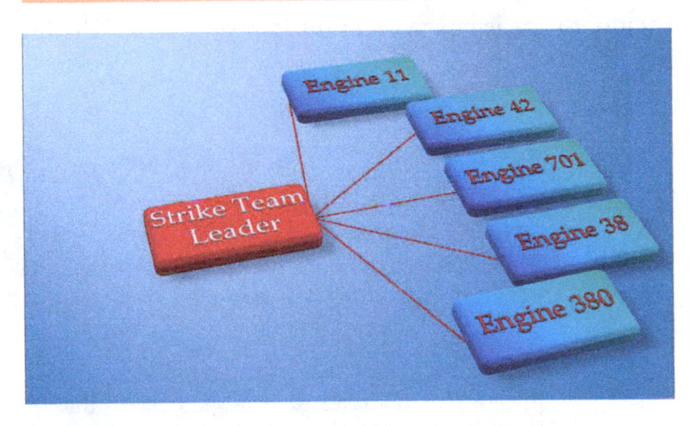

Figure 2-1 *Span of Control for a Strike Team of Engines.*

Incident Action Plans (IAP)

Every incident must have an Incident Action Plan (IAP). Small incidents may have a verbal action plan whereas a large incident will have a written IAP. The IAPs are often referred to as shift plans. **Incident Action Plans** must include a statement of objectives, organizational structure, tactical assignments to accomplish the objectives, weather report, fire behavior predictions, and any other supporting material. Two important concepts that are also included an incident action plans are a safety message prepared by the Safety Officer and a medical plan prepared by the Medical Unit Leader.

The IAP is critical to any incident. It is your responsibility as a firefighter to ensure you are familiar with the details of the plan and what your role is. Large fires may have 20 to 30 pages involved in their IAP. Whereas a small and short duration incident, like a small, one acre fire, does not require a 30 page IAP.

[2] National Wildfire Coordinating Group, *Firefighter Training S-130* (Boise: National Wildfire Coordination Group, 2003), 2.2.

> **INCIDENT ACTION PLAN**
>
> Every incident must have an Incident Action Plan. It can be either verbal or written.

When you are on a remote fire without the luxury of a copy machine, many incident commanders will prepare an incident action plan and deliver it to you verbally, but support it with visual references. The common technique is to draw the incident map with dry erase markers on a white or light colored vehicle. This provides the firefighters with a visual reference and allows them to ask questions and receive immediate feedback.

INCIDENT TYPES

Wildland fire incident types are categorized into five major categories. In relation to the incident types, a qualified Incident Commander will be appointed to manage the incident activities. Incident types are based on the complexity, duration, and/or size of the incident.

Incident Types

As mentioned above, there are five categories for incident management used in the ICS. Remember, the incident type is based on complexity, not size. The incident types are identified in the table below.

TABLE 2-1 *INCIDENT TYPES*

Incident Type	Certified Incident Commander	Complexity	Duration	Example	Example of Resources
Type 5	Type 5 IC	Minimal—lowest complexity level	Short—think 24 hours	• 1 acre fire • 2 car accident with no injuries	• 1 Fire Engine or • 1 police unit
Type 4	Type 4 IC	Increased from Type 5, but still low to moderate	Usually handled with 24–48 hours	• 10 acre grass fire with 2 days of suppression activity	• 2 Fire Engines • 1 Hand Crew
Type 3	Type 3 IC	• Extended attack • Logistical concerns are elevated • Numerous air resources • Values at risk	Over 24–48 hours	• 110 acre timber fire	• 6 Fire Engines • 3 Hand Crews • Air Coordinator • 5 Water tenders • Logistics is established • Operations Section is Established
Type 2	Type 2 IC	• Project fire • Evacuations • Structures threatened • Values at risk • Advanced air resources	Long duration event	2,400 acre timber fire threatening homes	• Full blown ICS structure
Type 1	Type 1 IC	• Project fire with the highest level of complexity or anticipated complexity • Major logistical considerations • Major air resources	Long duration event	16,000 acres for example, the Little Bear Fire	• Full blown ICS structure

Small Fire Organization

(Small scale, minimal complexity, short duration)

The National Wildfire Coordinating Group estimates that "90% of wildland fires are suppressed during the initial attack phase, with a small organization."[3] A majority of the fires you respond to will be small in nature. They will not require a large amount of resources to suppress them. In addition, logistical support will be minimal, with the fire resources on scene able to support themselves with food, water, and supplies that are on their assigned vehicle or equipment. Finally, most of these incidents will be finished in a short time period. It makes no sense to have a large management organization to deal with these small incidents. The beauty of the ICS is that it allows us flexibility to institute the appropriate management team or individuals to deal with a specific incident.

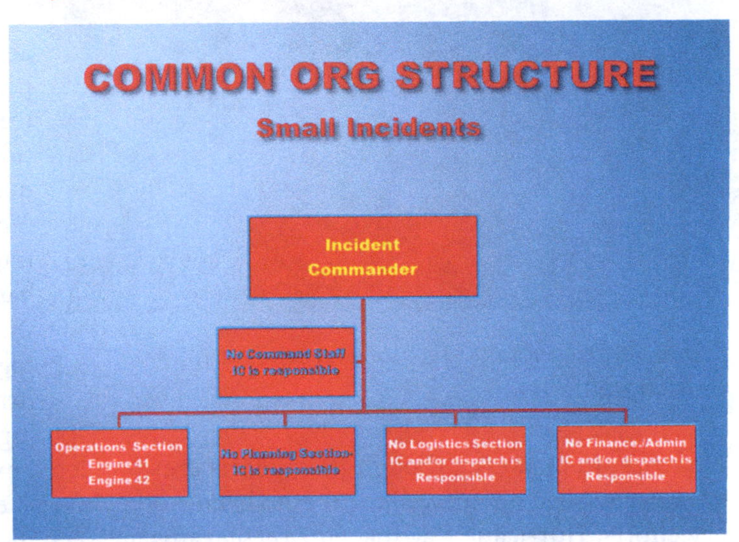

Figure 2-2 *Small Fire Organization.*

Large Fire Organization

As a fire increases in size and complexity, fire managers will utilize the ICS to ensure the needs of the incident are met. Twenty years ago, large fire organizations were determined based on the size of the fire. This is no longer the case. Larger fire organizations are now based on complexity, duration of the incident, and values at risk. The more complex an incident is, the bigger the management structure needed to successfully deal with it. Figure 2-3 illustrates a typical organizational structure for a large fire.

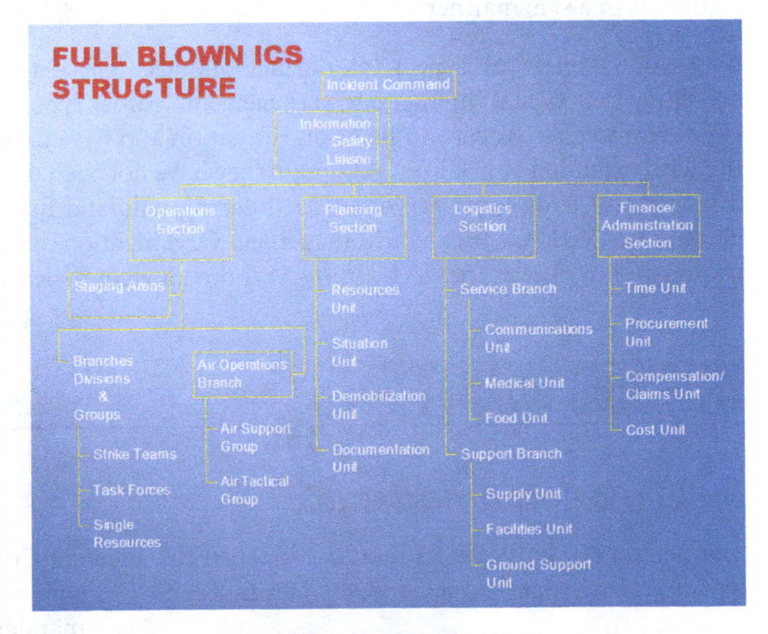

Figure 2-3 *Large Fire Organization. Courtesy of FEMA.*

[3] Ibid.

Interagency Management Teams

Courtesy of Brian Henington

Type 1 Team conducting a morning briefing in Washington, 2014.

Incident Interagency Management Teams are explained as "the incident commander and appropriate general and command staff personnel assigned to an incident."[4] These teams select members through an application process that is based on the individual's qualifications and experience. The team's members are from different agencies (thus interagency) and work together on a yearly basis. Because the teams are preassembled, the group can work effectively and efficiently together with each member aware of his or her specific roles and duties. These teams are highly essential to wildland fire suppression activities. In fact, even though these teams are considered experts in wildland fire management, they have been called on numerous times to handle other all-risk events, like the World Trade Center recovery efforts, Hurricane Katrina, etc. These teams are very successful in managing incidents because of their years of experience dealing with complex emergencies. No other organization, except the military, can develop a town in the middle of nowhere in a matter of days and coordinate massive amounts of resources in a safe manner.

There are sixteen (16) Type 1 Interagency Management Teams in the United States. In addition, there are thirty (37) Type 2 Interagency Management Teams. Type 3 Interagency Management Teams are also very common across certain parts on the nation and have demonstrated superior performance in dealing with extended attack fires. Although they have not been discussed in this chapter, there are two more management teams that are often used during wildland fire activities. They include Area Command Teams and National Incident Management Organizations. You will learn about these teams and their responsibilities in future ICS training. (We do not want to overwhelm you at this point of your training).

ICS JOB POSITONS

Incident Commander (IC)

The person in charge of an incident is called the Incident Commander (IC). He or she is responsible for all activities, personnel, and support functions of a particular incident. A successful incident commander is an effective leader who understands how to use the ICS in the best manner possible. He or she has to be able to implement command and control of the situation but not by micromanaging the activities of firefighters.

COMMAND AND CONTROL

The success of the ICS depends on the Incident Commander's ability to lead and implement the structure and command of the incident.

[4] Ibid., 105.

Job Duties

Courtesy of Glen Sveum

Type 4 Incident Commander conducting operational briefing.

- All activities are the responsibility of the incident commander. He or she may have to perform several roles on a small incident if authority is not delegated to another qualified individual.
- Quality incident commanders have the ability to delegate and communicate.
- Unsuccessful incident commanders try to manage all activities themselves or are micromanagers. It may be common for a Type 5 IC to be involved in the suppression of a fire, but Type 4 ICs and above are managers, not hands-on firefighters.
- Overall, the IC will ensure that safety for firefighters is prioritized above all other objectives, followed by the safety and general welfare of the public.
- The incident commader may not be the highest qualified individual on a wildland fire. A small fire does not require a qualified Type 1 Incident Commander to manage it. The incident would be managed by a Type 5 IC, who is lower qualified than the Type 1 IC.

© Karin Hildebrand Lau/Shutterstock.com

Information Officer talking to the media.

Command Staff

1. Deputy Incident Commander: If a deputy incident commander is required for an incident, then he or she must be as qualified as the person they work for. Deputy ICs are common in large fire organizations. Smaller incidents will require a deputy incident commander.
2. Information Officer: Information officers are responsible for interacting with the media, the incident commander, other fire personnel, and the public. Large incidents will require more than one information officer. On small incidents, the incident commander will function as the information officer.
3. Safety officer: The Federal Emergency Management Agency defines a safety officer as, "A member of the command staff responsible for monitoring and assessing safety hazards or unsafe situations, and for developing measures for ensuring personnel safety."[5] The safety officer is the only person who can override an incident commander's decision; this can only be done if it's based on safety concerns. Today, safety officers are more than people who check for earplugs or safety glasses. They strive to see the bigger picture and identify the hazards and risks

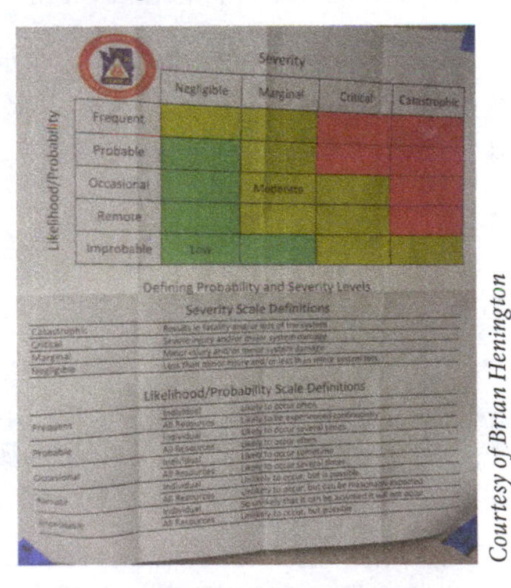

Courtesy of Brian Henington

Analysis of risk and hazard exposure completed by a safety officer.

[5] Federal Emergency Management Agency, *ICS Glossary* (Washington, D.C.: Department of Homeland Security. Federal Emergency Management Agency, 2008), 10.

firefighters are exposed to. They work hard to ensure that risk is reduced and firefighter safety considerations are prioritized at all times. Most safety officers have an extensive operations background that will help them determine if tactical decisions are correct. The safety officer is not involved in tactical decisions; he/she only provides insight as it relates to firefighter safety and risk management. There can be more than one safety officer on a fire.

4. Liaison Officer: A liaison officer is described as "A member of the Command Staff responsible for coordinating with representatives from cooperating and assisting agencies."[6] There can be more than one liaison officer on a fire.

It is important to note that small incidents do not require an assigned a safety officer, information officer, or liaison officer. The incident commander can perform all the duties of those positions because of the limited complexity and duration of the incident. Large scale or complex incidents will require these positions to ensure successful management, communication, and coordination.

The General Staff

The general staff reports directly to the incident or deputy incident commander. They include an Operations Section Chief, Planning Section Chief, Logistics Section Chief, and a Finance/Administration Section Chief. Again they are not needed on a small incident. Each section chief manages his or her responsible section. On a large-scale event, the operations section chief may have 1,000 firefighters under his or her control. Section chiefs will also use delegation to ensure that the span of control is in place at all times. No operations section chief can successfully manage 1,000 firefighters. He or she has to rely on other qualified overhead to help facilitate the needs of the incident and ensure tactical objectives are met.

Operations Section

As mentioned before, the operations section is managed by an Operations Section Chief on a Type 1 or Type 2 incident. A Division Supervisor or Group Supervisor can manage the operations section on a Type 3 Incident. On Type 4 or 5 incidents, the incident commander would be responsible for operational decisions and implementation of tactical objectives.

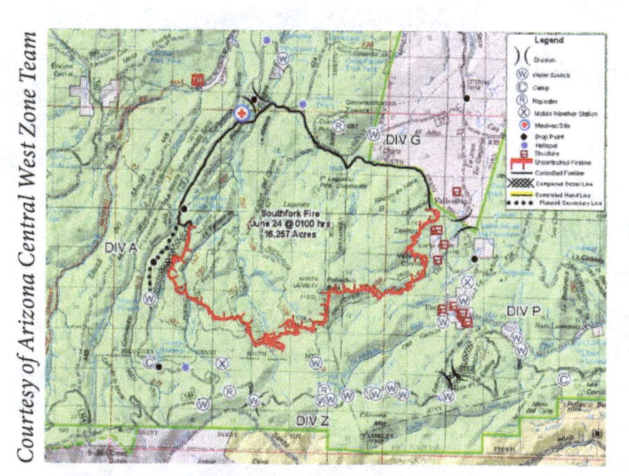

Courtesy of Arizona Central West Zone Team

This map illustrates divisions used on a wildfire. They are identified by DIV A, DIV G, DIV P and DIV Z.

Most first responders will begin their career in the operations section. They are responsible for filling sandbags to stop floods, digging fireline to stop an advancing wildland fire, performing the physical duties of search and rescue missions, or providing the technician activities for hazardous materials event.

Branches: Although not used on all fires, the operations section chief can activate branches if they feel the incident complexity is increasing and span of control is not in place. The person responsible for a branch is a Branch Director. Large incidents may have multiple branches.

Divisions: Divisions are highly important to wildland fire suppression activities. Divisions are separated by geographical responsibility (or from point A to point B). Divisions are used to divide incidents into specific areas

[6] Ibid., 7.

of operation. Additional divisions are created based on the span of control and complexity. Divisions grow as the incident grows and they can decrease as the incident decreases. Divisions are identified by letters using the phonetic alphabet. For example, Division A would be called Division Alpha. Division Z would be called Division Zulu. The person responsible for a division is a Division Supervisor.

Division: Geographical Responsibility
Group: Functional Responsibility

Group: A group is responsible for a function. It provides a specific function to the incident and can be used in one division or multiple divisions. The person responsible for a group is a Group Supervisor. An example of groups on wildland fires include structure protection group, initial attack group, or an evacuation group. Once it is done with its specific function, then the group will move to a new area of concern or could be dismantled based on the need of an incident.

Groups and divisions are at the same organizational level and report directly to the Operations Section Chief or Branch Directors. The Division Supervisor does not supervise the Group Supervisor, nor does a Group Supervisor supervise the Division Supervisor.

The following is an example of a division and group working together. When I worked for the state land office, we had district resource managers who were responsible for specific counties in the state of New Mexico. They managed all activities and monitored industry activities within their district. At our headquarters, we had specialized individuals who provided a specific function. An example of one was an archaeologist who was responsible for cultural resources. Once the district resource manager identified an issue, he or she would then call a supervisor (Operations Section Chief) requesting the support of the archaeologist. The archaeologist would then travel to the specific district, perform the required duties, and then leave. The district resource manager would remain in his or her assigned area. The archaeologist would then return back to headquarters and be ready for the next assignment.

Task force: A task force consists of three to seven resources of different kind and type. There may be several task forces working within the group or a division. The task force is managed by a Task Force Leader. A task force leader will report to a division supervisor or group supervisor. An example of a task force would be two fire engines, one water tender, and three dozers. This concept will be explained in more detail in Chapter Seven.

A task force in Oregon preparing to bed down at the end of their shift.

Strike Team: A strike team is managed by a Strike Team Leader. On paper, a strike team leader is not as qualified as a task force leader. A strike team has to have similar types and kinds of resources. An example of a strike team would be five (5) type six fire engines. They do not have to be from the same agency but they have to have the same capabilities. A strike team leader will report to a division supervisor or group supervisor or task force leader. Again, this concept will be explained in more detail in Chapter Seven.

The operations section is made up of **single resources**. Single resources can be equipment, fire engines, aircraft, or individual firefighters. If the ICS System is used correctly, fire managers will ensure the span

Courtesy of Brian Henington

The members of Engine 41 and Engine Boss Jim O'Leary are considered a single resource.

Courtesy of Brian Henington

Helicopter supporting fire activities.

of control is in place and group single resources into strike teams or task forces. Everybody works together to perform a specific duty under the ICS System and all elements focus on safety, communication, and coordination.

Structure of the 20 Person Hand Crew: A hand crew will consist of 18 to 20 firefighters. The firefighters will be managed by a crew boss or crew superintendent. Most 20-person hand crews will have one squad boss or advanced firefighter supervising four to six members of a squad. That squad could consist of chainsaw operators or Type 2 firefighters who are responsible for completing the physical requirements of the job. Effective hand crews will utilize the span of control to ensure the most effective and safe operation of their hand crew.

Air Operations: This branch of operations is responsible for the coordination of aircraft used on the fire. The activities they manage can be tactical application of aircraft or logistical support. The two categories of aircraft used on a wildland fire are fixed wing (airplanes, airtankers) and rotary wing (helicopters). Air activities are very complex and dangerous. Managing these activities requires extensive and effective communication, coordination, and cooperation.

Courtesy of Brian Henington

Incident map produced by the Planning Section.

Planning Section (Plans)

The planning section is "Responsible for the collection, evaluation, and dissemination of information related to the incident, and for the preparation and documentation of Incident Action Plans."[7] The planning section works the closest with the operations section. Most planning section personnel have an operational background, which helps facilitate the needs of the operations section. The plans section is made up of several units; however they will only be established if the complexity of the incident requires it. A small-scale fire does not require a plans section. The person responsible for the planning section is the Planning Section Chief.

The Units below are associated with the Planning Section:

- **Resources Unit:** Responsible for tracking and recording all resources committed to an incident. It is also responsible for anticipating additional resources.

[7] Ibid., 9.

- **Situation Unit:** FEMA describes the Situation Unit as, "responsible for the collection, organization, and analysis of incident status information, and for analysis of the situation as it progresses"[8]
- **Documentation Unit:** Maintains all documents related to the incident activities.
- **Demobilization Unit:** Plans for the release of personnel and equipment based on the need of the incident and the amount of days worked by each resource. You cannot leave an incident when you're ready to leave. You have to go through a formal demobilization process.
- **Field Observer:** Collects information from personal observations and supplies it to requesting personnel.
- **Other Planning Personnel:** Based on the need of the incident, the following positions can be activated to support the activities of the planning section.
    - » Status/Check-in Recorder
    - » Fire Effects Monitor
    - » Infrared Interpreter
    - » Display Processor
    - » Fire Behavior Analyst
    - » Strategic Operational Planner
    - » Long-Term Fire Analyst
    - » Geographic Information System Specialist
    - » Interagency Resource Representative
    - » Human Resource Specialist
    - » Incident Training Specialist

Logistics Section

© bikeriderlondon/Shutterstock.com

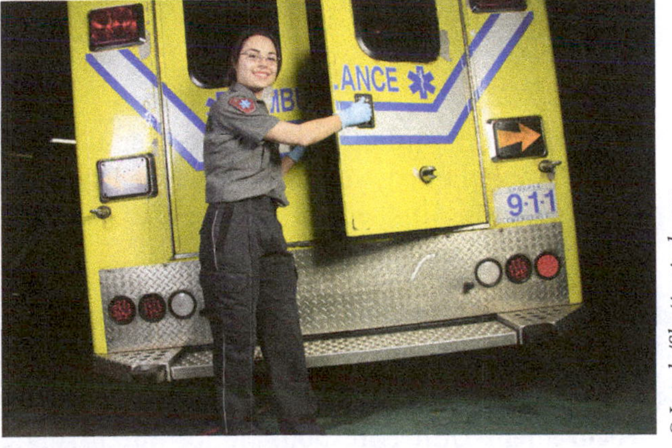

© Lopolo/Shutterstock.com

Entry-level firefighters should remember the function of logistics is to supply you with your needs. The logistics section is designed to facilitate the needs of the personnel supporting the emergency. They are essential and critical to effective management of the incident. The logistics section is managed by the Logistics Section Chief. On large fires, they can include two branches: Service Branch and the Support Branch. The service branch contains three units (if needed).

Communications Unit: This unit establishes the frequencies used on the fire. The communications unit will also ensure that working communications are established. If there are issues with communications, this unit will help facilitate those concerns. It will also provide batteries for handheld radios and fix radios that have issues. The communications unit also functions as an incident dispatch, as they will relay messages to the fireline from the incident command post or base.

Medical Unit: Provides medical needs to fire or emergency personnel. They do not provide needs to civilians. Injuries are common to wildland firefighters. They may be minimal like a sunburn or blisters,

---

[8] Ibid., 11.

or they may be severe. If a vehicle accident occurs during an evacuation, the medical response would fall under operations and not under the logistics section or medical unit. The person responsible for the medical unit is the Medical Unit Leader and he or she may have several people working in the unit that are typically qualified medical personnel. The leader is also responsible for developing a medical plan to be used by all personnel on the emergency.

Food Unit: The Food Unit is responsible for feeding firefighters and providing water and drinks. They will also provide ice as needed. On large incidents, the food unit will cater food and may be responsible for feeding 1,200 or more firefighters. They will provide a breakfast, sack lunch and a hot dinner.

© Tobias Arhelger/Shutterstock.com

Fire hose can be checked out at the Supply Unit.

The support branch of the logistics section involves the following:

Supply Unit: The Supply Unit is responsible for all equipment, tools, or fire clothing that has been damaged on the incident. They will also supply any other materials that are needed during the incident. All equipment is accountable; therefore, most materials require a formal checkout process. Some items must be returned before you leave the fire.

Courtesy of Brian Henington

Standard facilities used at incidents.

Facilities Unit: This unit is responsible for providing and managing fixed facilities at the incident. Examples include camp, base, and incident command post. They are also responsible for sleeping locations, showers, feeding areas, and sanitary facilities.

Ground Support Unit: This unit has several responsibilities but the most common are making repairs to vehicles, transporting firefighters to and from certain locations, or assigning rental vehicles. They will also provide fuel to fire equipment.

Finance/Administration

I always tell my young firefighters to keep the Finance/Administration Section happy and treat them fairly because they pay you! Of course, the Finance/Administration Section has far more responsibilities than paying us. This section can consist of the Time Unit, Procurement Unit, Compensation/Claims Unit and the Cost Unit.

Below is a brief summary of the responsibility of these units:

Time Unit is responsible for processing time of personnel assigned to the incident.

Procurement Unit is responsible for purchasing equipment and developing contracts for materials needed to support the incident.

Compensation/Claims Unit is responsible for documenting any accidents involving fire related vehicles, equipment, or firefighters. This unit is also responsible for ensuring the proper paperwork is completed for any such injury or accident.

Cost Unit will monitor costs on a daily basis and provide those costs to the Incident Commander and agency responsible for the fire. The resources and equipment needed to combat large wildfires are very expensive.

INCIDENT FACILITIES

Large incidents will require several facilities to support the emergency response activities. Smaller incidents will not necessarily require all of these facilities. The most common facilities used on wildland fires are briefly identified below:

- **Incident Command Post (ICP)**: FEMA describes the Incident Command Post (ICP) as "the field location at which the primary tactical-level, on-scene incident command functions are performed. The ICP may be collocated with the incident base or other incident facilities and is normally identified by a green rotating or flashing light."[9] The ICP on small fires will be located wherever the IC is located; it can be mobile. On a large fire, the ICP will be an established location, regardless of the Incident Commander's location.

Courtesy of Brian Henington

- **Staging Areas**: Staging areas are part of the operations section. Established for tactical considerations, they are designed to provide a location for resources to gather, organize, and be ready for tactical assignment. Staging areas require a three-minute response time if the forces are activated.
- **Base**: The **Incident Base** is described as, "The location at which primary Logistics functions for an incident are coordinated and administered. (Incident name or other designator will be added to the term Base.) The Incident Command Post may be collocated with the base."[10]

© Barry Blackburn/Shutterstock.com

Resources awaiting tactical assignment in a staging area. Three-minute dispatch required.

- **Camp**: A camp is where firefighters sleep, eat, and shower. The camp is usually located away from the base, but they can be close to each other (within walking distance).

Courtesy of Brian Henington

Courtesy of Brian Henington

[9] Ibid., 5.

[10] Ibid., 2.

- **Spike Camp:** A location separate from the camp that is established based on logistical concerns on the incident. Spike camps will normally not have as much support as a camp and often lack facilities.
- **Helibase:** The National Wildfire Coordinating Group identifies a helibase as "the main location within the general incident area for parking, fueling, maintenance, and loading of helicopters. It is usually located at or near the incident base."[11] Rural airports are often used as helibases.
- **Helispots:** A temporary landing or take off area for helicopter operations. They are also used to provide a geographical reference for firefighters. Helispots are identified as: H-1, H-2, H-48, etc.

Courtesy of Brian Henington

Spike Camps are established for logistical concerns. In the case of this spike camp, two crews were camped closer to the fire for quick access.

© Wollertz/Shutterstock.com

Refueling of helicopters occurs at a helibase.

© ChiccoDodiFC/Shutterstock.com

Helispots are temporary landing locations for helicopters.

SUMMARY

The Incident Command System can be confusing. It contains a lot of information and concepts that are introduced over a short time frame. As an entry-level firefighter, you do not have memorize and understand every concept. What you must understand is the concept of chain of command, span of control, and how your job fulfils the objectives of this management system. As you gain more experience and advance your training, the concepts in this chapter will become more familiar to you. Your first large fire assignment may be overwhelming. Count on your supervisor to assist you with the functions of ICS and how those functions will benefit you. You should further understand what sections or units of the ICS can support your medical needs, your supply needs, or vehicle needs such as refueling. This system is highly functional if it is used correctly. Once you understand the system, you will better understand the amount of effort, coordination, and cooperation needed to manage emergency and wildland fire incidents.

[11] National Wildfire Coordinating Group, *Glossary*, 101.

KNOWLEDGE ASSESSMENT

1. What section of the ICS conducts tactical operations to carry out the plan and directs all tactical resources?
2. True or False: The Incident Commander is always the most qualified individual on a wildland fire.
3. On a wildland fire, the Medical Unit is in what section?
4. True or False: Wildland firefighters, structural firefighters, and law enforcement officers can be organized under the same incident command system.
5. True or False: Divisions and Groups are at the same organizational level and can work together on an incident.
6. True or False: The Incident Command System is not designed for small incidents.
7. True or False: Deputies must always be as qualified as the person for whom they work.
8. In what section and unit can you locate hand tools, sleeping bags, safety gear, and 1 ½" hose?
9. In ICS, communication is in ________________.
10. The five major management activities (positions) around which the ICS is organized are:
11. Span of control may vary from ________________ to ________________, with a reporting element of ________________ subordinates to ________________ supervisor.

EXERCISES

1. The General Staff consists of:
2. What are the three (3) major activities (positions) of the command staff?
3. Division Supervisors have ________________ responsibility.
4. Group Supervisors have ________________ responsibility.
5. Identify the common ICS responsibilities of all emergency responders.

BIBLIOGRAPHY

Federal Emergency Management Agency. *ICS Glossary.* Washington, D.C.: Department of Homeland Security, Federal Emergency Management Agency, 2008.

National Wildfire Coordinating Group. *Firefighter Training S-130.* Boise: National Wildfire Coordinating Group, 2003.

National Wildfire Coordinating Group. *Glossary of Wildland Fire Terminology.* Boise: National Wildfire Coordinating Group, 2014.

CHAPTER 3

WILDLAND FIRE BEHAVIOR BASICS AND COMMON TERMINOLOGY

Learning Outcomes

- Explain the fire triangle.
- Discuss the heat transfer process and how it impacts wildland fires.
- Identify the nine parts of a wildland fire.
- Define key terminology used during wildland fire management activities.
- Explain the importance of anchor points and why they must be used during suppression activities.

Courtesy of Willie Lucero

Key Terms

Anchor Point
Burn Period
Chain
Crown Fire
Fire Triangle
Heat Transfer Process
Mop-up
Spot Fire
Torching

OVERVIEW

The ability to understand and comprehend fire behavior is an essential and necessary proficiency of every firefighter supporting wildland fire activities. Fire behavior directly impacts our safety, the tactics we use, and will influence how long our efforts are needed to fully extinguish a fire. Over the past 20 years, many of my students have found fire behavior to be confusing or boring. They would rather study the tactics, equipment, and techniques we use. I would challenge this misconception by saying that there is no way that you can be a safe and effective wildland firefighter without first understanding how fire behaves under specific environmental influences. You must know what situations are dangerous, which circumstances present demanding challenges, and what conditions will impact your personal safety. If this is the career for you, you must fully commit yourself to studying and understanding the influences that drive and influence wildland fire behavior.

What is Fire and Fire Behavior?

It is important to understand what the definition of fire is regardless if you are a structural or wildland firefighter. According to a published and well-respected fire science professor, Dr. James Lally, "Fire is a set of chemical reactions that produce heat and light."[1] The Merriam-Webster dictionary further defines fire as "an occurrence in which something burns: the destruction of something (such as a building or a forest) by fire."

Although fire is often viewed as destructive, it is also important to note that it has an important and valuable role in our society. It is essential for warming our homes, cooking, maintaining healthy forests, and it is used in the production of consumer goods and materials. It is not always the enemy, (when used in a controlled manner), but in our profession it should be viewed as a living organism that is capable of destruction and harm to both the firefighter and the public. At all times we must respect what fire is capable of. Always treat every fire as if it has the potential to be extreme. This approach will ensure that you focus on safety and do not take chances or risks that would have negative results.

Fire behavior impacts our safety and should dictate our strategy and tactics. The more intense and extreme the fire behaves, the more caution and concern we must have. The National Wildfire Coordinating Group classifies fire behavior as "the manner in which a fire reacts to the influences of fuel, weather, and topography."[2] We should fight every fire as aggressively as we can; however, only after all safety measures are in place and implemented. Be a student of fire! Understand the characteristics of fire behavior and base your actions on current and expected fire behavior. If you do this, you will have a successful career as a wildland firefighter.

The **fire triangle** represents the three elements needed to support combustion. The three elements must be in the right proportion and must be subject to an ignition source in order for a fire to occur. There is no fire if the three elements are not present or if there is not an ignition source.

---

[1] James, Lally, PhD, "What is Fire: The Definition and Exploration of Fire Behavior," in *Fire Behavior and Combustion* (August 28, 2014).

[2] National Wildfire Coordinating Group, *Glossary of Wildland Fire Terminology PMS 205* (Boise: National Wildfire Coordinating Group, 2014), 72.

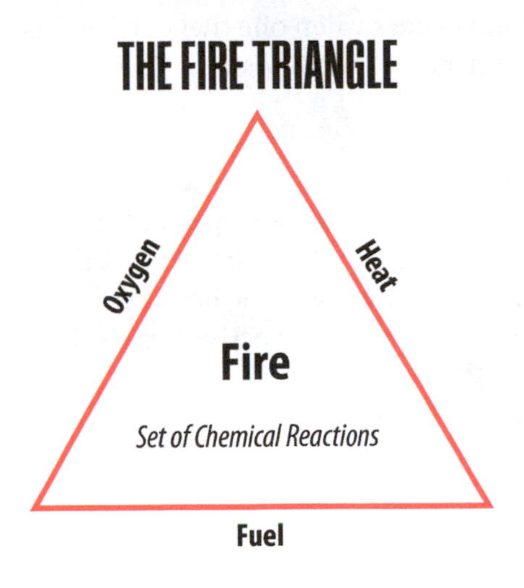

Figure 3-1 *Fire Triangle.*

- **Heat:** In wildland fire, heat is highly influenced by the sun. The hotter the day or time of year, then the more potential for a fire to occur. This is the main reason why wildland fires occur during the hotter parts of the year and not during the winter. In addition, fire activity is usually the greatest between 2:00 pm to 6:00 pm (referred to as the **Burn Period**), which is typically the hottest part of the day. As firefighters, there are techniques to control heat that is produced by a fire, but there limited measures to control heat produced by the sun.
- **Oxygen:** Fire is a living organism. It must have oxygen. In a wildland environment, oxygen is more than available, with winds bringing unlimited supplies of it. As firefighters, there are limited mechanisms available to control oxygen during the active burning stages of a fire. Available oxygen is primarily controlled by Mother Nature.
- **Fuel:** Fuel is any material that is available to burn. Out of all three elements of the fire triangle, fuel is the component that we can mitigate and control the most. The main reason wildland firefighters construct fireline is to remove available fuel from the advancing fire. In theory, the more fuel available, the larger, more intense fires we will have.

HEAT TRANSFER PROCESS

There are three components involved in the **heat transfer process**. They include radiation, convection, and conduction. This process is important to wildland fire because it captures how heat moves from one object (or fuel) to another object (or fuel), thus supporting continuous burning or combustion. As mentioned earlier, fire produces heat. That heat will move vertical or side-to-side. As heating occurs and touches fuel that has yet to burn, then that fuel loses moisture and becomes more ready or available to burn. We are primarily concerned with radiant and convective heat in our environment.

Convection occurs when hot air moves upward or vertically. It is a concern to us because it can preheat fuels that are directly above the burning object. Convection also contains the very dangerous component of smoke as well as the chemicals created from combustion. In addition, convective heat will also travel up steep slopes and preheat and dry fuels above the fire. Once the flaming front reaches the preheated and dried fuels, the chemical reactions of fire occur at exceptional speed and intensity.

Radiation is the movement of thermal energy through the air. Radiant heat moves in all directions. The best example of radiant heat is the heat the sun produces. On a wildland fire, radiant heat can impact surrounding and adjacent fuels of the burning object. Once they reach a certain temperature, they can ignite, which can create very fast moving and hard-to-control fires.

Conduction is the least of our concerns on a wildland fire. Conduction occurs when one fuel particle has direct contact with another fuel particle and heats it to a level that can support combustion.

Teaching Illustration: A good example of the heat transfer process is a fire in your fireplace. The heat you feel sitting on your couch is radiant heat. It moves in all directions. The more logs you put in the fireplace (conductive heat is involved), the warmer you will get. We put chimneys on houses to allow convective heat to escape. You probably have always heard that we lose a majority of the heat from our fires out of the chimney. This is true. The reason we do not want to capture the convective heat is that it contains the smoke and chemicals produced by the fire.

WILDLAND FIRE TERMINOLOGY

Parts of a Fire

The following definitions depict the nine (9) common fire behavior terms used in fire suppression activities.

1. **Head of a Fire:** The head of the fire is the side of the fire that is moving the fastest. It typically indicates the direction the fire is moving, especially when influenced by wind or topography. The head is considered the most dangerous part of the fire. We should avoid attacking a fire in front of the head. (This concept will be explained in later chapters.)
2. **Point of Origin:** The exact location where ignition occurred and sustained combustion. It is where the fire started. Once identified, the point of origin should be protected from destruction and maintained for investigation purposes.
3. **Rear of a Fire:** The opposite of the fire head, it usually is at the point of origin, especially if the fire is influenced by the wind. The rear is the safest location to start attacking a fire because it has burned the longest and intensity is less than at the head. The rear of the fire will also move slower, especially when influenced by wind and/or topography. The rear of the fire is commonly referred to as the "Heel of a Fire."
4. **Flank of a Fire:** The flanks of a fire are included in the fire perimeter and typically run parallel to the fire spread. The flanks of a fire are determined by standing at the rear of the fire looking towards the head. Once determined, the flanks become geographical reference points for firefighters. For example: "Engine 42, can you meet me on the right flank?" or "Smokey Bear Hotshots, can you relocate to the left flank to begin firing operations?" Remember, we do not have streets or other reference points to pinpoint suppression activities. Why are the flanks important? Tactically, firefighters move up the flanks of the fire to conduct tactical operations. Although the flanks are considered not to be as dangerous as the head, they still can be very active and have dangerous fire behavior.
5. **Fire Perimeter:** The complete outside edge of a fire.

PARTS OF A FIRE

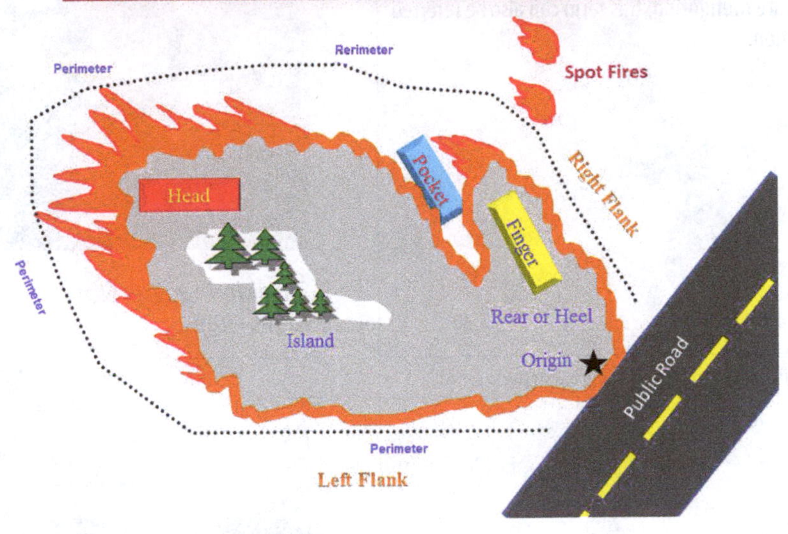

Figure 3-2 *The Parts of a Wildland Fire.*

6. **Fingers:** According to the National Wildfire Coordinating Group (NWCG), a finger of a fire is defined as "the long narrow extensions of a fire projecting from the main body."[3] There are many factors (such as available fuel or wind direction) that can cause a fire to move away from the main fire. Fingers can be dangerous because there is a greater opportunity for firefighters to become outflanked or trapped if the fingers burn together or back to the main fire. Tactically, fingers should be cut off and included in the main body of the fire.
7. **Pockets:** Pockets are the area(s) between fingers. They contain unburned fuel which impacts suppression efforts.
8. **Islands:** Islands occur within the main body of the fire. For whatever reason (fuel changes, decrease in fire behavior, fuel moisture), the area remains largely unburned or completely unburned.
9. **Spot Fire:** Spot fires occur outside or away from the main fire. They can start from wind blow fire embers (fire brands) or from rolling debris on steep slopes. Spot fires have the potential to become very dangerous and they are usually associated with extreme fire behavior. Spot fires are classified into two categories: long-range spotting and short-range spotting.

Common Fire Behavior Terms

Wildland fire is dominated by terminology. Understanding the terms is important and should be considered a top priority for all of us, regardless of our experience or qualification level. For the purpose of this book, we will introduce you to the common fire behavior terms that entry-level firefighters should be aware of and understand prior to fighting any fire. The following information provides definitions for these core concepts.

---

[3] Ibid., 71.

Smoldering: Smoldering is considered combustion without any visible flame. In structure firefighting, this term can also be referred to as glowing combustion.

Courtesy of Brian Henington

Smoldering Fire: Notice there is no active flame.

Creeping Fire: Fire with low flame and minimal activity. Fire spread is slow.

Courtesy of Brian Henington

Minimal fire activity.

Running Fire: Fire activity is increasing. The head of the fire is well defined and visible.

© James Bostian

Well-defined head with increased intensities.

Spotting: Usually associated with extreme fire behavior. Spotting causes spot fires. Fire brands or embers are transported in the wind or convection column of the fire. If the embers are still on fire and land in available fuel, then new fires can ignite. Spotting can also be influenced by gravity—causing fire brands to roll down hill on steep slopes.

Courtesy of Brian Henington

This spot fire occurred ¼ mile away from the main fire.

(continued)

Torching: The consumption or burning of tree or brush foliage. Usually involves one tree or bush or a small group of trees/brush. Starts at the bottom of the fuel species and moves vertically to the top. Considered a phase of a crown fire—passive crown fire. The result of torching can be fire brands, which start spot fires. Firefighter note: Torching is a step needed for a crown fire to occur.	 Courtesy of Brian Henington *Torching—single tree becomes involved in fire from bottom to the top.*
Crown Fire: A fire that advances from the top of trees/brush to the tops of adjacent trees/brush. Produces extreme burning intensity and fire activity. Presents the highest complexity and danger to firefighters. Considered extremely dangerous.	 Courtesy of Brian Henington *Crown fire.*
Flare Up: Sudden increase in fire behavior, either the spread of a fire or intensity. Usually a short-lived event.	 Courtesy of Brian Henington *Increase of fire activity; the flare up in the picture lasted a couple of minutes.*

(continued)

Firewhirl: Firewhirls occur under extreme burning conditions. They are similar to a dust devil and are created when hot air and gas rise vertically and begin to spin creating a vortex. Firewhirls vary in size with reports of some occurring up to 500 feet in diameter.

Courtesy of Brian Henington

Firewhirl created in by a prescribed fire. The bottom of the vortex is full of fire.

Backing Fire: A fire spreading into the wind or downslope. The intensity of a backing fire is usually minimal and the spread is slow.

Courtesy of Brian Henington

Flames are backing into the wind. Minimal spread and activity.

Flaming Front: The National Wildfire Coordinating Group defines a flaming front as, "That zone of a moving fire where the combustion is primarily flaming." The heavier the fuel burning, the deeper or wider the flaming front. If lighter fuels are burning, the narrower the flaming front.[4]

© James Bostian

Well-defined flaming front.

Blowup: Fire activity is burning at the highest rates and intensity. "Sudden increase in fireline intensity or rate of spread of a fire sufficient to preclude direct control or to upset existing suppression plans. Often accompanied by violent convection and may have other characteristics of a fire storm."[5]

© Tom Reichner/Shutterstock.com

[4] National Wildfire Coordinating Group, *Glosary*, 83.

[5] Ibid., 36.

Additional Fire Suppression Terms

The following terms are essential knowledge for every firefighter participating in wildland fire activities.

The **anchor point** of a fire is the location in which fire suppression activities begin. An anchor point is defined by National Wildfire Coordinating Group as "An advantageous location, usually a barrier to fire spread, from which to start constructing a fireline. The anchor point is used to minimize the chance of being outflanked by the fire while the line is being constructed."[6] We cannot stress how important an anchor point is. You may have multiple anchor points on any given fire, but the essential concept here is you must have anchor points. This ensures we will not be outflanked by a fire.

Teaching Illustration: Have you ever been fishing for crappie? Crappie are a warm water fish (and very delicious) that run in large schools. When you find the school, you will usually catch multiple fish. To ensure you stay above the school of crappie, you drop an anchor at that particular location. The anchor ensures you are locked in place and do not move away from the fish. Anchor points lock you in place. They ensure that fire will not move around or outflank you. By establishing a quality anchor point, you are ensuring that you can safely and effectively attack the fire.

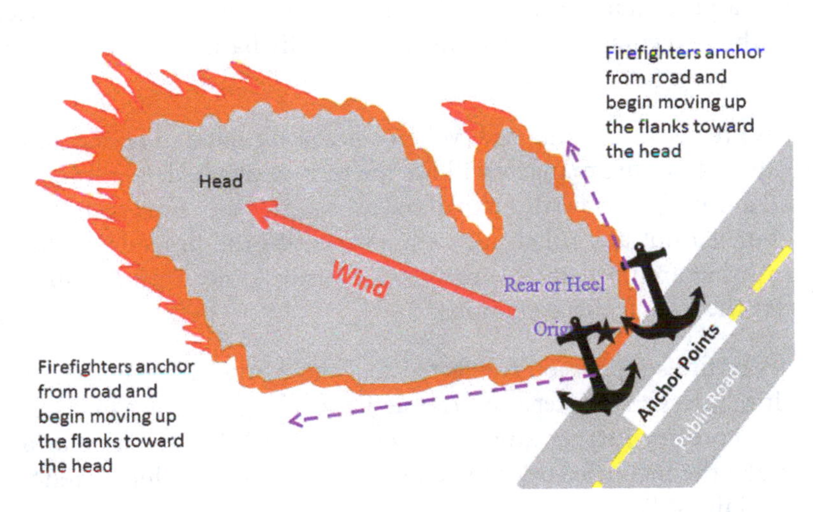

Figure 3-3: *Illustration of Anchor Points.*

Another important term is a **control line**. The United States Forest Service defines a control line as "All built or natural fire barriers and treated fire edge used to control a fire".[7] Control lines can be natural, such as a river or overgrazed pasture, or can be man-made, like a road. They are also considered any fireline that is constructed during suppression activities. A **fireline** occurs during the life of the fire. It is constructed by firefighters using hand tools, equipment, machines, explosives, and/or wetting or chemical agents.

[6] National Wildfire Coordinating Group, *Introduction to Wildland Fire Behavior S-190 Student Workbook* (Boise: National Wildfire Coordinating Group, 2006), 1.8.

[7] United States Forest Service, *Fire Terminology*, accessed February 2, 2015, http://www.fs.fed.us/nwacfire/home/terminology.html#C.

Courtesy of Brian Henington

The road would be considered a control line.

Courtesy of Brian Henington

Dozer constructing a fireline.

The ultimate goal of fire suppression activities is to contain and control the fire. Containing or **containment** of a fire identifies the progress firefighters have made in their efforts to suppress a fire. It is defined as "the status of a wildfire suppression action signifying that a control line has been completed around the fire, and any associated spot fires, which can reasonably be expected to stop the fire's spread."[8] It is usually identified in a percentage. For example, fire managers will update the accomplishments of firefighters by stating the percentage of containment on a daily basis; an example would be: the fire is 10% contained.

The term **controlled** is greater than contained. When a fire is expected to no longer move and the lines are protected and secured, than fire mangers will use the term controlled. Controlled is "the completion of control line around a fire…"[9] It is further explained as, "… any spot fires therefrom, and any interior islands to be saved; burned out any unburned area adjacent to the fire side of the control lines; and cool down all hot spots that are immediate threats to the control line, until the lines can reasonably be expected to hold under the foreseeable conditions."[10]

One of the final stages of fire suppression is referred to as **mop-up**. In laymen terms, mop-up involves putting the fire out. It requires many steps and can demand a large number of firefighters to complete. The Alaska Division of Forestry further explains this concept as "after the fire has been controlled, all actions required to make the fire 'safe,' prior to being called out. This includes trenching, falling snags, and checking all control lines".[11]

Finally, the last basic term a wildland firefighter must be familiar with is the term **chain**. A chain is a measurement tool. The term was adopted by our community from early surveyors. A chain represents 66 feet. We use the term to communicate distances such as chains of fireline that need to be constructed, how many chains the fire is moving in an hour, or the objectives call for mop-up activities two chains from the fire perimeter. The chart below provides a breakdown of chains as it relates to feet, yards and miles.

---

[8] National Wildfire Coordinating Group, *Glossary*, 49.

[9] Ibid., 50

[10] Ibid., 50

[11] Alaska Department of Natural Resources, Division of Forestry, *Glossary of Fire Terms*, accessed February 2, 2015, http://forestry.alaska.gov/fire/glossary.htm.

TABLE 3-1 *CHAINS RELATED TO FEET, YARDS, AND MILES*

Chains	Feet	Yards	Mile(s)
1	66	22	
2	132	44	
20	1,320	440	1/4
40	2,640	880	1/2
80	5,280	1,760	1

AVAILABLE WILDLAND FIRE BEHAVIOR TRAINING

Additional and more advanced fire behavior training for firefighters is available through the National Wildfire Coordinating Group. I highly recommend that you continue improving your knowledge and comprehension of wildland fire behavior by participating in one or more of these courses (Note—most NWCG classes require prerequisites for advanced training. Those courses are identified as 200 level courses and above).

- *S-190 Introduction to Wildland Fire Behavior:*
  This course was developed to provide entry level wildland and structural firefighters with the knowledge necessary to understand wildland fire behavior and the three primary environmental factors that influence the start and spread of a wildland fire.
- *S-290: Intermediate Wildland Fire Behavior:*
  This class expands on the basic concepts of wildland fire behavior by providing a more in-depth and focused study on the three environmental factors that influence wildland fire behavior. The class is required training for an Advanced Firefighter or a Fire Effects Monitor. It is also the foundation for any single resource boss performing job duties on a wildland fire.
- *S-390 Introduction to Wildland Fire Behavior Calculations:*
  This class is designed to fulfill the job requirements of a strike team leader, task force leader, and/or Type 3 incident commander. The class concentrates on wildland fire behavior calculations by using tables and monograms and provides a detailed comprehension of the factors related to wildland fire behavior.
- *S-490 Advanced Wildland Fire Behavior Calculations:*
  This advanced wildland fire behavior course is intended to provide a scientific approach to determine inputs for wildland fire behavior. The course focuses on interpretations of model outputs as well as prepares students to predict fire perimeter growth based on weather predictions and knowledge of fuels and topography.
- *S-590 Advanced Wildland Fire Behavior Interpretation:*
  This course is highly advanced as it was developed to prepare future Fire Behavior Analysts and Long Term Fire Analysts positions used in wildland fire suppression and fire use organizations.

SUMMARY

Fire behavior has a direct impact on firefighter safety. It is your responsibility to improve your knowledge level related to fire behavior. Learn as much as you can about this subject and apply it throughout your career as a firefighter. This is the first chapter related to wildland fire behavior. The following Chapters (*Chapter 4: The Fire Behavior Triangle* and *Chapter 5: Critical Fire Conditions and Monitoring Techniques*) will provide an introduction to the fire behavior triangle and the influence that the environment has on fire behavior and firefighter safety.

KNOWLEDGE ASSESSMENT

1. What is the definition of fire?
2. What time of the day is considered the burn period?
3. What are the three elements of the fire triangle?
4. Out of the three elements of the fire triangle, which one can be controlled or mitigated the most by wildland firefighters?
5. Name and briefly define the nine parts of a wildland fire. Which part is considered the most dangerous? Which part is where the point of origin typically occurs?
6. Which fire behavior term is considered the most dangerous and presents the most complexity to fire suppression activities?
7. Define the term contained and controlled.
8. How many chains are in a mile? How many chains would be in 2.5 miles?
9. List the three components of the heat transfer process. Which two are we most concerned with as it relates to fire suppression activities?

EXERCISES

1. On a piece of paper, draw and identify the nine parts of a fire. In addition to the parts of the fire include at least two anchor points (identify them by a large A).
2. Try an experiment. Take a candle (ensure you are somewhere safe and the candle flame will not touch available fuel). Light the candle. Observe the convective heat lifting vertically above the candle. Now feel for heat next to the candle (at the sides). This would be radiant heat. Which heat (above or to the side) is more intense?

BIBLIOGRAPHY

Alaska Department of Natural Resources, Division of Forestry. *Glossary of Fire Terms.* http://forestry.alaska.gov/fire/glossary.htm.

Lally, James PhD. "What is Fire: The Definition and Exploration of Fire Behavior" in *Fire Behavior and Combustion*, August 28, 2014

Merriam-Webster Dictionary. *Definition of Fire.* http://www.merriam-webster.com/dictionary/fire.

National Wildfire Coordinating Group. *Introduction to Wildland Fire Behavior S-190 Student Workbook NFES 2901.* Boise: National Wildfire Coordinating Group, 2006.

National Wildfire Coordinating Group. *Glossary of Wildland Fire Terminology PMS 205.* Boise: National Wildfire Coordinating Group, 2014.

United States Forest Service. *Fire Terminology.* http://www.fs.fed.us/nwacfire/home/terminology.html#C.

CHAPTER 4

THE FIRE BEHAVIOR TRIANGLE

Learning Outcomes

- Explain how topography features influence wildland fires.
- Identify the six (6) wildland fire fuel types.
- Explain the concept of available fuel for combustion.
- Describe ladder fuels and the laddering effect.
- Examine the impact temperature has on fire behavior.
- Compare stable and unstable atmospheric conditions.

Courtesy of Brian Henington

Key Terms

Available Fuel
Barriers
Box Canyons
Fire Behavior Triangle
Fuel
Slope
Terrain
Topography
Weather
Wildland Fuel

OVERVIEW

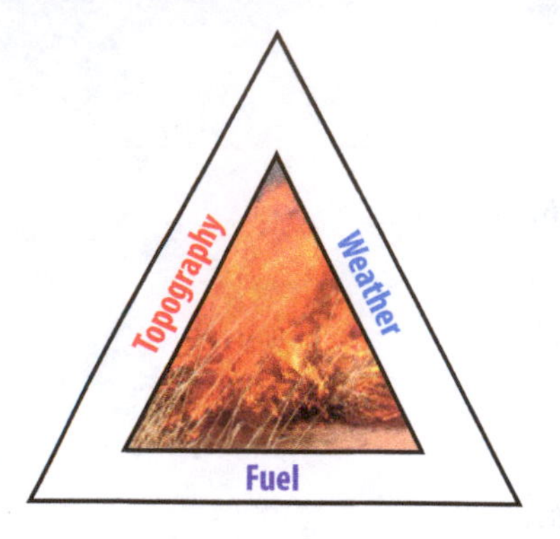

Figure 4-1 *The Fire Behavior Triangle.*

Wildland fire behavior is highly impacted by three environmental factors: topography, fuel, and weather. These factors, known as the **Fire Behavior Triangle**, represent certain elements or events that contribute to fire behavior and firefighter safety. It is our responsibility to become familiar with the components of the Fire Behavior Triangle to ensure that our tactical decisions are based on the best information available. After completion of this chapter, you should be familiar with the different factors of the triangle and be able to explain their influence(s) on fire behavior and your personal safety.

TOPOGRAPHIC FEATURES AND THEIR INFLUENCE ON FIRE BEHAVIOR

As firefighters, we have to be aware and considerate of topography features. The National Wildfire Coordinating Group defines **topography** as, "the configuration of the earth's surface including relief in the position of its natural and manmade features.[1] These features have the ability to highly influence fire spread and activity. Our concern is based on the impact it has on our safety. For the focus of this chapter, we will concentrate on ten topographical features that influence fire behavior. The key features include: aspect, slope, terrain or shape of country, box canyons, narrow canyons, wide canyons, ridges, saddles, elevation, and barriers.

Courtesy of Brian Henington

Aspect is the direction a slope is facing the sun. The aspect of a slope must be considered by firefighters because of its impacts on fuel and fuel moisture. Typically, south, southwest, or southeast aspects have the most exposure from the sun; thus the fuel is more susceptible to burn. North aspects typically have more moisture, larger trees, and higher humidity; therefore they are less apt to burn. (However, years of drought impacting the western United States have challenged this theory on north aspects). As firefighters, we are most concerned with the aspects that have the greatest exposure from the sun. Aspects are identified as south, north, west, east or southeast, northeast, etc.

[1] National Wildfire Coordinating Group, *Introduction to Wildland Fire Behavior S-190* (Boise: National Wildfire Coordinating Group, 2006), 2A.3.

Another critical consideration fire managers must consider is the topographical feature called a **slope**. A slope is defined as "the angle of incline on a hillside."[2] Slope has a major impact on fire behavior. The steeper the slope, the faster a fire will travel. The reason this occurs is based on several factors:

- Radiant and convective heat travel up the slope, which causes preheating and drying of fuels above the flaming front.
- Flames are closer to unburned fuel above them, causing fire spread to occur very rapidly.

The location of the ignition point is also important. If lightning ignites a fire close to the top of a hill, then the amount of available fuel and preheating would be limited. If an ignition occurs at the bottom of the slope, more fuel is available for the fire and preheating will have greater impact.

Gravity also plays an important role on slopes. Steep slopes present the opportunity for burning material to roll downhill, which may cause new ignitions or endanger firefighters. Think of it this way: You have always been told not to try to outrun a bear uphill because bears run uphill faster than they do downhill. The same holds true for fire. Fire will spread uphill with great velocity and energy. As firefighters, we want to avoid being above a fire on steep slopes. All the dangers of smoke, chemicals released by the fire and the fire intensity are heading up hill. Don't be above the danger!

Fire behavior is highly influenced by steep slopes.

The **terrain** or **shape of the country** will also impact our safety and have major influences on fire behavior. Terrain can be explained as the physical features of a specific parcel of land. Terrain has the ability to influence fire behavior by directing wind movement, fire spread, intensity, and speed. Terrain features include saddles, canyons, ridges, etc. The information below describes major terrain features that influence fire behavior and impact firefighter safety.

Box Canyon: A box canyon can be best described as a canyon that has one adequate ingress route. The egress route would also be the same route. One way in, the same way out! Box canyons are usually very steep at the head (or top) of the canyon and on the sides. A box canyon is dangerous because it funnels wind and fire behavior. Similar to the way a chimney in your house functions, a box canyon will draw fire from the bottom of the canyon to the top with great velocity. Don't go into a box canyon! Avoid them! Throughout history of wildland firefighting, box canyons have been involved in far too many firefighter fatalities.

Courtesy of Brian Henington

Photograph of a box canyon after a fire has burned. Notice how steep the top of the canyon and sides are. No firefighters entered this area, it was allowed to burn.

[1] National Fire Protection Association, Fire Wise Communities, *Understanding Fire Behavior*, accessed February 3, 2015, http://learningcenter.firewise.org/Firefighter-Safety/1-9.php.

Teaching Illustration—Firefighter Safety Note: Use your imagination. Picture a river running down a very steep canyon. Pressure is increased every step it travels downhill. Box canyons would function in the opposite manner. Every step the fire moves up the canyon, the more intensity the fire gains.

Narrow Canyon: Narrow canyons influence upslope wind movement and preheating on adjacent slopes. If fire is introduced in a narrow canyon, wind influences will likely cause elevated fire activity. Fire burning in narrow canyons can also cause preheating on opposite slopes which increases the available fuel for combustion. Spotting is another critical factor in narrow canyons. Tactical operations should avoid narrow canyons because of the elevated risk and exposure to firefighters.

Wide Canyon: Wide canyons impact fire behavior through wind movement. Prevailing wind direction and movement can be impacted in wide canyons.

Ridges: Ridges are terrain features that indicate the crest of a hill. They are typically narrow and long, and run to the summit of the mountain. The major influence ridges have on fire behavior is wind movement on opposite sides of the ridges. Be concerned with an ignition occurring on the opposite side of the ridge (in the unburned area) and running uphill. Larger ridges may provide a tactical advantage to fire suppression activities.

Saddle: The best way to describe a saddle is where two hills join or come together. Saddles funnel fire! We should avoid saddles for tactical activities because of the funneling effect they have on fire.

Elevation: NWCG defines elevation as "the height of the terrain above the mean sea level, usually expressed in feet."[3] Elevation will impact fuel types and seasonal drying as well as testing the firefighters' cardio conditioning.

[1] National Wildfire Coordinating Group, *Wildland Fire Behavior*, 2A.6.

Courtesy of Brian Henington

The river can function as a natural barrier that can be used strategically to contain fire spread.

Courtesy of Brian Henington

A man-made barrier, like this gravel road, can also be used to contain fire spread.

Barriers: A barrier can be used to obstruct the spread of a fire. Barriers can be either natural or man-made. Examples of natural barriers include: rivers, lakes, overgrazed or drought-stricken pastures, rock-slides, or previously burned areas. Man-made barriers may include roads, ditch banks, highways, fuel breaks, etc.

WILDLAND FUELS

We discussed how topographical features can impact fire behavior and the safety of firefighters. However, the driving factor of fire behavior is available fuel. Across the United States, we have different fuel types and fuel species. Fuels react to fire in different manners. As a firefighter you need to be familiar with the fuel types in your jurisdictional area. You should know the common fuel species by name. I once asked a firefighter what type of tree was burning. His response was "I don't know; it just burns!" That is not the proper attitude to have as a firefighter. If you are a structure firefighter, you understand how the fuel within a structure impacts fire behavior. You are familiar with the materials used to make furniture, carpet, building materials, as so forth. Each type of fuel will impact or add to the structure fire. The same holds true for wildland fires. You should be familiar with fuel species and types because they will behave differently if consumed by fire.

The goal of this section is to provide information related to the basic components of fuel as it relates to their impact on fire behavior. We will introduce you to the six standard fuel types and discuss other characteristics of fuel you should be aware of.

Teaching Note: Many untrained people will claim that fuel cannot be modified or changed in any manner. This is not true! Think back to the fire triangle introduced in *Chapter 3*. The one element we can control the most as firefighters is fuel. We construct fireline to remove fuel away from the fire. Agencies across the United States are spending millions of dollars to mitigate fuel conditions to reduce the potential for catastrophic fire. We use prescribed fire to reduce fuel accumulations. Fuel can be mitigated, if we have enough time and resources to do so.

What is Fuel?

Fuel is essential to a fire. The National Wildfire Coordinating Group defines fuel as any burnable material.[4] They further explain **wildland fuel** as any live and/or dead plant material that is available to burn. The energy produced by a fire is directly related to the fuel that is consumed. In theory, the more fuel, the more potential a fire has. However, not all fuel is available. **Available fuel** describes fuel that will burn and sustain combustion. A green football field would not be considered available because it would not support or sustain combustion. A field of three-foot high dry grass will support combustion and should be considered available.

Wildland fuel is determined based on the primary carrier of the fire that typically occurs within the surface area. We do not use the crowns of trees or tree species themselves to determine fuel types because they are not considered the primary carrier of the fire. When a fire elevates into the crowns of the trees we consider this to be "fire in the third dimension." When you arrive on scene of a fire, you should be looking at the surface fuels and determine which fuel type is carrying the fire. (Some fires may have multiple fuel types.)

Fuel Types

Wildland fire managers classify wildland fuel into six types. They include 1) grass, 2) grass/shrub, 3) shrub, 4) timber/understory, 5) timber litter, and 6) slash-blowdown.

1. **Grass**

   Key characteristics:

   » Burns very fast (in tall grass) and is highly influenced by wind speed and direction.

   » Short grass will have minimal fire activity with low to moderate movement.

   » Green grass will be difficult to sustain combustion.

   » Very common throughout the United States.

   » Varies in size and species. Some grass may be six feet tall.

   » Precipitation and relative humidity will impact fuel moisture.

Courtesy of Brian Henington

Tall grass—expect increased fire behavior.

Courtesy of Brian Henington

Fire behavior in tall grass.

[4] National Wildfire Coordinating Group, *Glossary of Wildland Fire Terminology PMS 205* (Boise: National Wildfire Coordinating Group, 2014), 88.

Short grass.

Courtesy of Brian Henington

Typical fire behavior in short grass.

2. **Grass/Shrub**

Key characteristics:

» Primary carrier of the fire will be intermixed grass with shrubs and/or forbs.

» Common in semi-arid climates (high desert) and the plains areas of the United States.

» Fire spread and intensity will be high, based on the intermixed fuels.

» Direct exposure by the sun will cause drying of the fuels.

Courtesy of Brian Henington

Grass/shrub fuel type. Sage intermixed with grass.

Courtesy of Brian Henington

Fire burning in grass/shrub fuel type. Increased fire activity.

3. **Shrub (or Brush)**

Key characteristics:

» Primary carrier of the fire will be both live and dead foliage, branches and leaves of the shrubs.

» Under the worst burning conditions, this fuel type can be extremely volatile with high intensities and spread.

» Very flammable shrub fuel species:

- Sagebrush in the Great Basin states.
- Palmetto/gallberry in southeastern United States.
- Chaparral in Arizona and California.

Courtesy of Brian Henington

Primary carrier of the fire would be brush.

© Tom Fawls/Shutterstock.com

Brush burning in Florida. The primary carrier of the fire in this photo is the brush.

4. **Timber/Understory**

Consists of shrubs, small trees, grass, and other fuel that grows on the surface underneath a timber or tree overstory.

Key characteristics:

» Primary carrier of the fire is forest litter with herbaceous fuels or brush.

» Found throughout the United States.

» Ladder fuels are a major concern in this fuel type. (Ladder fuels will be explained later in this chapter.)

Courtesy of Brian Henington

The primary carrier of the fire would be the fuels growing under the understory.

Courtesy of Brian Henington

The understory is providing the energy for the fire shown.

5. Timber Litter

Includes the needles, leaves, small branches, bark, twigs, pinecones, etc. from the trees growing above.

Key characteristics:

» Very common throughout mountain regions in the United States.

» Main driving force of fires occurring in mountain areas in New Mexico/Arizona.

» Includes duff, which is defined as the layer of decomposing organic materials lying below the litter layer of freshly fallen twigs, needles, and leaves and immediately above the mineral soil.[5]

The man carrier of the fire would be pine needles, small branches, pinecones, etc.

Active fire behavior at night in timber litter.

Firefighter burning out timber litter.

6. Slash/Blowdown

Includes the debris or remains left after natural events or human activity. This type of fuel typically has low to moderate fire spread, but can burn very hot. Debris or remains may include stumps, logs, branches, broken tree tops, Examples of activities that create debris include:

» Logging

» Thinning

» Wind/snow events

» Road building

» Clearing or rights-of-ways

[5] National Wildfire Coordinating Group, *Glossary*, 63.

Courtesy of Brian Henington

Small, scattered slash. Notice the red cast of the needles. It is an indication of little to no fuel moisture. Expect extreme temperatures when ignited.

Courtesy of Brian Henington

Lopped and scattered slash. This activity is the result of a thinning project.

Courtesy of Brian Henington

Piles of slash. This type of pile is against most state laws; however, you may have a fire threatening or burning in an area like this.

Courtesy of Brian Henington

A look into a slash pile burning. High burning temperatures produced.

Courtesy of Brian Henington

Slash piles burning. Extreme temperatures but low rate of spread.

Courtesy of Brian Henington

Fully-involved slash pile.

Fuel Loading

Fuel loading is defined as "the amount of fuel expressed quantitatively in terms of weight of fuel per unit area (tons per acre)."6 In laymen's terms, fuel loading is the amount of fuel in a given area that is available for combustion, if it is susceptible to burn. Take for example: A football field that has been mowed would have less fuel loading as the same football field that was allowed to grow to one foot in height. Of course this example would not support combustion unless the grass was without high fuel moisture.

Courtesy of Brian Henington

The grass would be considered low fuel loading.

© James Bostian

The grass would be considered high fuel loading.

Fuel Size Classes

Wildland fire managers have divided fuels into four distinctive class sizes. The class sizes use diameter as the primary factor for classification. Fuel size classes are important in monitoring and predicting fire behavior. They are also used in the preparation of prescribed fire plans and modeling. These four classes are classified based on the amount of time it would take the specific class to loss or gain moisture. The smaller or fine fuels lose and gain moisture the fastest—within one hour—whereas the thousand-hour class would take a thousand hours to lose complete moisture.

TABLE 4-1 *FUEL SIZE CLASSES*

Size Class	Diameter	Visual Reference
1-hour	0 to 1/4 inch	Courtesy of Brian Henington

(continued)

[6] Ibid, 89.

Size Class	Diameter	Visual Reference
10-hour	¼ to 1 inch	 Courtesy of Brian Henington
100-hour	1–3 inches	 Courtesy of Brian Henington
1,000-hour	3–8 inches	 Courtesy of Brian Henington

Fuel Arrangement

Fuel arrangement is the "manner in which fuels are spread over a certain area."[7] Fuel arrangement is separated into two categories: horizontal continuity and vertical arrangement. Horizontal continuity is further separated into two subcategories: uniform and patchy. Uniform or continuous fuels are fuels that have direct or immediate contact with adjoining fuels. Fire activity is typically higher in uniform fuels because of the opportunity for fire to spread easier from fuel to fuel. In addition, preheating and drying also play an important role with uniform fuels. The football field example would be considered uniform or continuous.

The second subcategory of horizontal continuity is patchy fuels. Patchy fuels do not have immediate contact with the adjoining fuels. Patchy fuels can be found in any ecosystem. Fire behavior would be greatly reduced in this arrangement, with isolated burning and limited fire spread.

[7] National Wildfire Coordinating Group, *Wildland Fire Behavior*, 2B.8.

Courtesy of Brian Henington

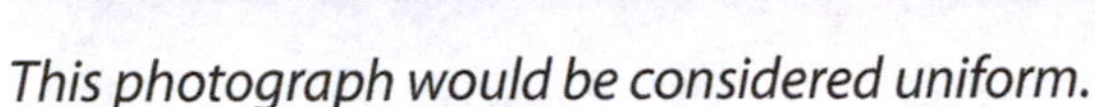

This photograph would be considered uniform.

Courtesy of Brian Henington

This photograph illustrates patchy fuels. Notice the amount of dirt between each fuel.

The second category of fuel arrangement is vertical arrangement. Vertical arrangement classifies "fuels above ground and their vertical continuity, which influences fire reaching various levels or vegetation strata."[8] Vertical arrangement has three categories.

1. **Ground fuels** are the combustible materials that occur beneath the surface or your feet. They can include any organic material, duff, roots, some stumps, or buried logs.
2. **Surface fuels** (from your feet to your eyes) include living or dead materials on the surface that can support combustion. Grass, timber litter, timber understory, ladder fuels, logs, and small brush occur within the surface. The surface fuels are what drives a fire or where the initial energy of a fire is generated.
3. **Aerial Fuels** occur above your eyes. They are located in the upper canopies of both trees and tall brush/shrubs. They can encompass dead and/or alive fuel. They may also include moss, snags, and tree branches. This category is where crown fires occur. It is important to note that not all fires reach the aerial fuels, but when they do, increased and extreme fire behavior should be expected.

Vertical Arrangement of Fuels

[8] National Wildfire Coordinating Group, *Glossary*, 181.

Ladder Fuels

Ladder fuels are considered any small trees and/or brush that grow in the surface area. They are considered ladder fuels because they allow fire to travel from the surface to the tops or canopies of trees. This concept is often called the laddering effect.

Courtesy of Brian Henington

The small ponderosa trees/saplings are considered ladder fuels.

Courtesy of Brian Henington

Fire has laddered from the surface to the aerial fuels.

How does it occur? Lightning strikes a tree. The electricity travels down the tree and ignites available fuels such as timber litter or grass. The fire begins to smolder than increases to creeping. The creeping fire begins to consume more timber litter and grass, increasing the energy of the fire. As burning continues, more fuels become involved. A small ponderosa pine sapling catches on fire. The fire slowly travels up the tree, consuming all the needles. There is a six-foot-tall ponderosa pine tree growing right above the sapling. The fire from the sapling catches the six-foot-tall tree on fire. The consumption of this tree then catches a 15-foot-tall ponderosa tree on fire. This tree becomes involved rather quickly because of the heat below it. Fire travels particularly fast through this tree until it comes in contact with the mature ponderosa pine. The mature tree now begins to torch. This is the laddering effect.

Figure 4-2

In theory, crown fires will not occur if the surface fire does not have the opportunity to ladder from each sublevel to sublevel until it reaches the crowns of mature or dominant trees.

Fuel Moisture

Fuel moisture can be described as the amount of water in any given fuel. The higher the moisture (the green football field), the limited amount of fire activity. The lack of fuel moisture (a football field that has not been watered for two months) would be more susceptible to combustion. Fuel moisture is measured in a percentage based on the oven-dry weight of the fuel. A low percentage would indicate low fuel moisture. The opposite would be true for high fuel moisture. Finally, light fuels (fine fuels) lose and gain fuel moisture greater than heavy fuels and are highly impacted by precipitation and relative humidity.

WEATHER'S IMPACT ON WILDLAND FIRE

Weather is created by changes in the atmosphere and is usually short-term that impacts day-to-day activities. The impact weather has on fire cannot be understated. Out of the three elements of the fire behavior triangle, weather is the most variable, unpredictable, and dynamic. Weather events have been associated with too many firefighter fatalities. On the other hand, weather may benefit our suppression efforts by bringing moisture or high relative humidity. The National Wildfire Coordinating Group demands that firefighters understand and monitor weather. "The risk involved in fire suppression can be reduced if firefighters and fire managers pay attention and understand weather conditions that impact fire behavior."[9] As a firefighter, you need to be familiar with weather events and the impacts they have on your personal safety. Weather episodes typically have visual indicators. Learn the indicators and apply them to your firefighting activities.

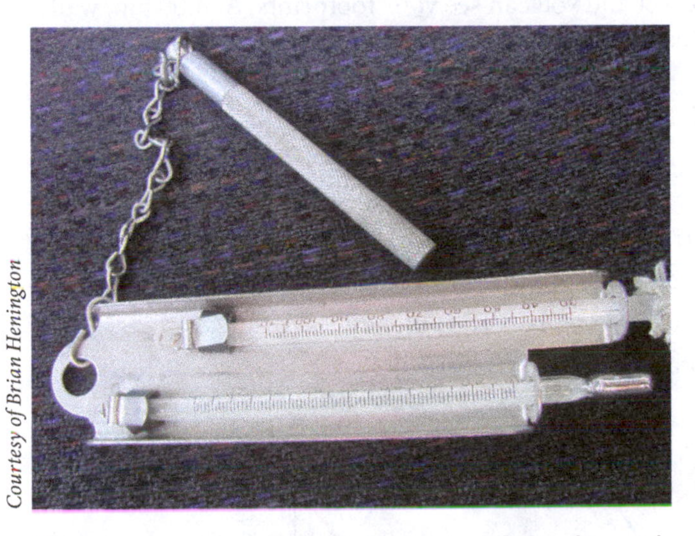
Courtesy of Brian Henington

The instrument is a sling psychrometer. The tool is used to measure both dry bulb and wet bulb temperatures on the fireline.

Temperature

Temperature provides a measurement of hot and cold. It is used to determine the heat or coolness of a substance or object. In wildland fire, we measure the temperature of air and apply it to our decision making. Basically, the hotter the day, the more opportunity for increased fire behavior. The colder the day, then fire behavior should be minimal. Temperature will vary depending on aspect, time of the day, and/or elevation changes. Because of this, it is important that firefighters constantly take temperature readings throughout the operational period. Notice that changes and monitoring trends can ensure that we are making the appropriate decisions on the fireline. Temperature is measured in either Fahrenheit or Celsius.

Relative Humidity

According the National Weather Service, "relative humidity is a ratio, expressed in percent, of the amount of atmospheric moisture present relative to the amount that would be present if the air were saturated."[10] In laymen's terms, relative humidity (RH) is the amount of moisture in the air. For our concerns, the lower the moisture, or RH, in the air the more and extreme fire behavior we could have. The higher the RH, the lower the fire behavior (this is not always the case depending on which geographical area you fight fire in).

---

[9] National Wildfire Coordinating Group, *Wildland Fire Behavior*, 2C.3.

[10] National Weather Service, *Definition of Relative Humidity*, accessed February 3, 2015, http://graphical.weather.gov/definitions/defineRH.html.

Relative humidity has a major impact on fuel moisture. The higher the relative humidity, then the greater the opportunity for fine or light fuels to gain moisture. On the other hand, the lower the relative humidity, the more opportunity for the atmosphere to take moisture away from the fine or light fuels.

The Relationship between Relative Humidity and Temperature

Temperature and relative humidity have an indirect or inverse relationship. When temperature is at its highest (midafternoon), then relative humidity would be at the lowest. As one increases, the other will decrease. At night, temperature will be at its lowest; thus, relative humidity would be at its highest. Because of this relationship, fuel moisture will also be impacted. At night, fuel moisture in fine fuels will be its highest throughout a 24-hour time period due to the higher amount of relative humidity present in the air. It is important that firefighters monitor temperature and relative humidity throughout the day to monitor or predict spikes in fire behavior.

Teaching Illustration: Try this. Next time you head to school (at 7 am) on a warm day, walk across some grass. You will notice that your shoes are wet and you can see your footprints. At 4:00 pm, walk across the same grass. Are your feet wet? No, this is an example of relative humidity's relationship with temperature.

Remember the following core concepts of the relationship between temperature and relative humidity:

- Relative humidity will be at its highest during the nighttime or during moisture events.
- Temperature will be at its highest during midafternoon (the Burn Period).
- Relative humidity will be at its lowest when temperature is at its highest.
- Relative humidity is measured in a percentage (0-100%).

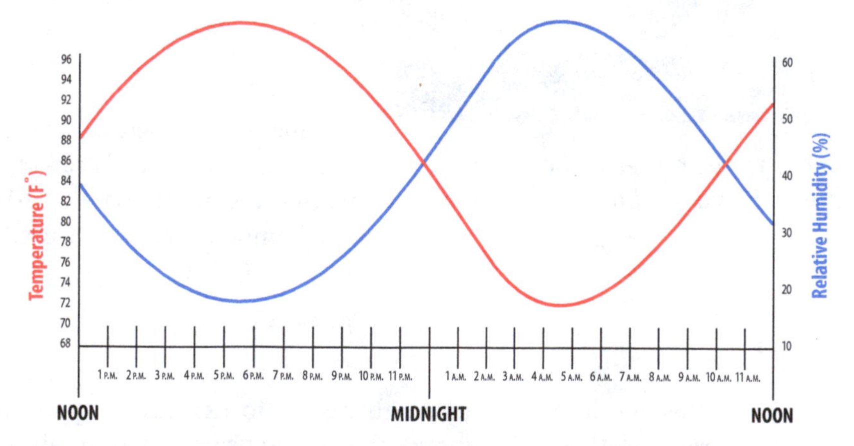

Figure 4-3 *Relationship between Temperature/Relative Humidity.*

Precipitation

Precipitation can directly impact current or expected fire behavior. Precipitation may appear in the form of rain, drizzle, snow, sleet, fog, or hail. Precipitation is formed in the atmosphere and falls to the earth's surface. The impact precipitation will have on a fire is directly related to the amount of moisture present.

A slow drizzle or wetting rain will have more impact than a short, very hard rain. Fine fuels are more impacted by precipitation than heavy fuels.

The more precipitation we receive, then, in theory, the limited fire activity we will have. I remember one spring and summer when I was a seasonal firefighter that it rained almost every day. We only had a few fires during that year. In 1994, we had very little rain in the spring and summer and hardly any snow pack during the previous winter. My fire engine responded to more than 100 fires that year.

Atmospheric Conditions

What is occurring in the atmosphere? Why should firefighters be aware or concerned? Atmospheric stability identifies the degree of vertical motion occurring in the atmosphere. This degree will either be enhanced or suppressed. Enhanced stability is referred to as unstable atmospheric conditions. A suppressed atmospheric condition is called a stable atmosphere. Fire burning under stable conditions will behave very differently than a fire burning under unstable conditions.

A stable atmosphere struggles against upward movement. An inversion is a common indicator of a stable atmosphere. An ongoing fire burning under stable conditions will not have vertical development or motion. The smoke will become suppressed and the conditions will be stagnant (like a pond that is not moving). Fire behavior is limited and intensity is minimal.

- Visual indicators: You are under a stable atmosphere if you can see one or more of the following:
  - Clouds in layers
  - Smoke column has limited rise and drifts apart
  - Fog or inversions
  - Constant, steady winds
- An inversion occurs under stable conditions. An inversion is cold air on the surface that is trapped by warm air above it.
  - Under an inversion fire behavior will be limited.
  - Visibility will be poor or limited because of smoke being trapped by the warmer air above it.
  - Inversions usually occur in the early morning.
  - Once an inversion lifts, movement in the atmosphere occurs; thus the fire has more available oxygen. Expect increased fire behavior once an inversion lifts.

Inversion layer—stable atmospheric conditions.

- Firefighters across the United States must be aware of inversion types and how they will impact fire behavior. Some common inversions are:
  - » Nighttime inversion: Occurs during the nighttime. Poor visibility. Lifts as the day warms.
  - » Thermal belts: Occurs in mountainous regions. Warmer air is trapped between cold air above and below it. Increased fire activity will occur in the thermal belt (usually mid-slope) because of increased temperatures.
  - » Subsidence inversion: This is a large-scale event that is associated with high pressure systems. Expect low relative humidity and warm, dry weather.
- Monitoring Inversions: Firefighters can monitor trends to determine when an inversion will lift. Some indications that an inversion is starting to lift are:
  - » Temperature increase
  - » Wind direction change and/or wind speed increase
  - » Relative humidity begins to decrease.

An unstable atmosphere has upward movement or vertical lift. Firefighters should expect to have increased fire behavior under unstable conditions because of the vertical movement of air. Unstable atmospheres contribute to increased fire behavior because:

- Convection column will lift vertically and produces stronger in drafts and convective lift.
- Firebrands will be transported vertically in the smoke column. At a certain point, they will fall to the surface which has the potential to start new ignitions.
- Firewhirls are more common.
- Surface winds will be stronger and gustier than they would be under stable conditions.

Courtesy of Brian Henington

Extreme fire behavior during an unstable atmosphere.

Firefighters should monitor visual indicators to determine if the fire they are fighting is burning under unstable conditions. Some indicators include:

- Vertical development of clouds—reaching great heights in the atmosphere.
- Cumulus clouds
- Good visibility
- Dust devils or firewhirls
- Gusty surface winds—blowing dust

Courtesy of Brian Henington

Smoke column lifts vertically and to great heights during unstable conditions.

Wind

Wind can be extremely dangerous to firefighters. Of the weather elements mentioned above, wind has the greatest impact on fire behavior. Strong winds have been associated with numerous firefighter fatalities. Wind can be variable over time and space which presents the challenge of predicting it.

Courtesy of Brian Henington

NWCG defines wind as "The horizontal movement of air relative to the surface of the earth."[11] Unlike convective columns that lift vertically, wind will move along the surface on the earth in a horizontal manner. To be effective firefighters, we must monitor wind conditions, (including speed and/or direction) at all times.

Wind has several impacts on wildland fire behavior. Below is a list of the most relevant.

1. Increases the amount of oxygen available for combustion.
2. Influences fire spread and direction.
3. Transports fire brands in front of the fire, which can cause spot fires.
4. Causes the convection column to shear or bend which causes preheating of available fuels in front of the fire.
5. Prior to fire activity, high wind events will dry fuel, which will make them more susceptible to burn if an ignition were to occur.

Furthermore, we must establish how the direction of wind is determined. Wind direction is determined by the direction the wind is blowing from—not the direction the wind is going to. A southwest wind would indicate the wind is blowing from the southwest, whereas an east wind would indicate the wind is coming from the east.

Question: If a fire was experiencing southwest winds, what direction would the head of your fire be traveling? Answer: Southwest winds would push the fire to the northeast.

There are several wind systems that you must be familiar with as a firefighter.

- A general wind is a large-scale event that is caused by the movement of high and low pressure systems. This system may impact fire behavior if the winds are strong enough and do not have terrain obstacles.
- A local wind is influenced the most by terrain. These winds are lower level in the atmosphere and have great impact on fire behavior. Local winds are generated by differences in temperature and pressure. Impact to fire behavior is a given if the winds are influential enough. Local winds include several types. They are briefly identified below:

[11] National Wildfire Coordinating Group, *Glossary*, 187.

» Upslope winds typically occur during the day. These winds are strongest during the day and when temperatures increase. The steeper the slope, the higher the wind speed. These events can be consistent to a specific area and can dominant prevailing winds if a major wind event is not present. Average wind speed is between 3 mph and 8 mph. The winds also have major influence on the movement of fire upslope. This is another illustration of why you should not be above an active fire during daytime burning activities.

- Downslope winds are most common at night. They occur when cooler air along mountain-tops begins to sink. Downslope winds are not as powerful or intense as upslope winds. Average downslope winds range from 2 mph to 5 mph. The change from upslope to downslope winds may be so subtle that it is not obvious. Taking weather on the hour will help firefighters determine when the wind change is occurring.
- Up-valley winds are larger scale events than upslope winds. They occur with the heating of the day, and are greatest in the late afternoon. Average wind speeds vary from 10 mph to 15 mph.
- Down-valley winds also have temperature changes at play. As the air cools during the night, down-valley winds become more prevalent (especially after midnight). The wind speed is less than up-valley winds with ranges occurring between 5 mph and 10 mph. Note that this is still enough wind to influence fire spread.

Sea and Land Breezes

Finally, we need to consider the influences that sea and land breezes have on fire activity. These winds are associated with large bodies of water. A sea breeze is a daytime breeze that can reach up to 30 mph. They are common in California and the Pacific Northwest. During the night, the wind will shift, becoming a land breeze. These winds will not be as strong as the sea breeze. If you live in an area that is impacted by sea and land breezes, you should learn their patterns and characteristics because they will impact fire behavior.

SUMMARY

You have just covered the elements involved in the fire behavior triangle. These environmental influences on fire behavior can be confusing and overwhelming to entry-level firefighters. I want to stress that the intent of this chapter is to introduce you to the concepts that influence fire behavior and your safety as a firefighter. Throughout your career you will continue to expand, develop, and increase your knowledge base in these areas. What you should take from this chapter are the key elements that will impact your safety.

You should understand how dangerous box canyons are. Don't go into them! If a supervisor asks you to conduct activities in a box canyon, then say something. You have the right to refuse the assignment based on safety concerns. Steep slopes influence fire behavior. There is no reason to be above a fast moving fire on steep slopes. In addition, fuel is the key to the start and spread of a fire. Understand that the more fuel available for consumption, the larger and more potential for extreme fire behavior. The western United States has been stricken with drought over the last several years. Trees and brush that appear alive are in fact weakened and more susceptible to burn. Our fires are bigger and more devastating and fuel plays a major factor. Get familiar with your fuel types and fuel species in your area and understand how the burn. If you understand these core concepts your career as a firefighter will be one that allows you to return unharmed at the end of each assignment.

KNOWLEDGE ASSESSMENT

1. Which aspect presents the greatest opportunity for fire spread? Which aspect (under normal conditions) would have the least opportunity for fire behavior?
2. Why do fires travel up slopes faster than they do down slope?
3. Why are box canyons dangerous to firefighters?
4. Horizontal continuity plays an important role in fire spread. Name the two categories. Out of the two categories, which one would have higher fire spread?
5. List the three categories of vertical arrangement.
6. Explain how ladder fuels allow a fire to travel from the surface to the crowns of trees or brush.
7. What is temperature and why is it important for firefighters to monitor?
8. What time of day would you expect temperature to be at the highest? When would relative humidity be at its lowest?
9. What is the relationship between temperature and relative humidity?
10. What event or occurrence is usually associated with stable atmospheric conditions?
11. How do you determine the direction of wind?
12. If a fire was experiencing southwest winds, what direction would the head of your fire be traveling?
13. List four local winds.

EXERCISES

1. Research dangerous wildland fuel types in your geographical area. Why are they dangerous? Have they been involved in firefighter fatalities in your area?

BIBLIOGRAPHY

National Fire Protection Association, Fire Wise Communities. *Understanding Fire Behavior.* http://learningcenter.firewise.org/Firefighter-Safety/1-9.php.

National Weather Service. *Definition of Relative Humidity.* http://graphical.weather.gov/definitions/defineRH.html.

National Wildfire Coordinating Group. *Glossary of Wildland Fire Terminology PMS 205.* Boise: National Wildfire Coordinating Group, 2014.

National Wildfire Coordinating Group. Introduction to *Wildland Fire Behavior, S-190.* Boise: National Wildfire Coordinating Group, 2006.

CHAPTER 5

CRITICAL FIRE WEATHER AND MONITORING TECHNIQUES

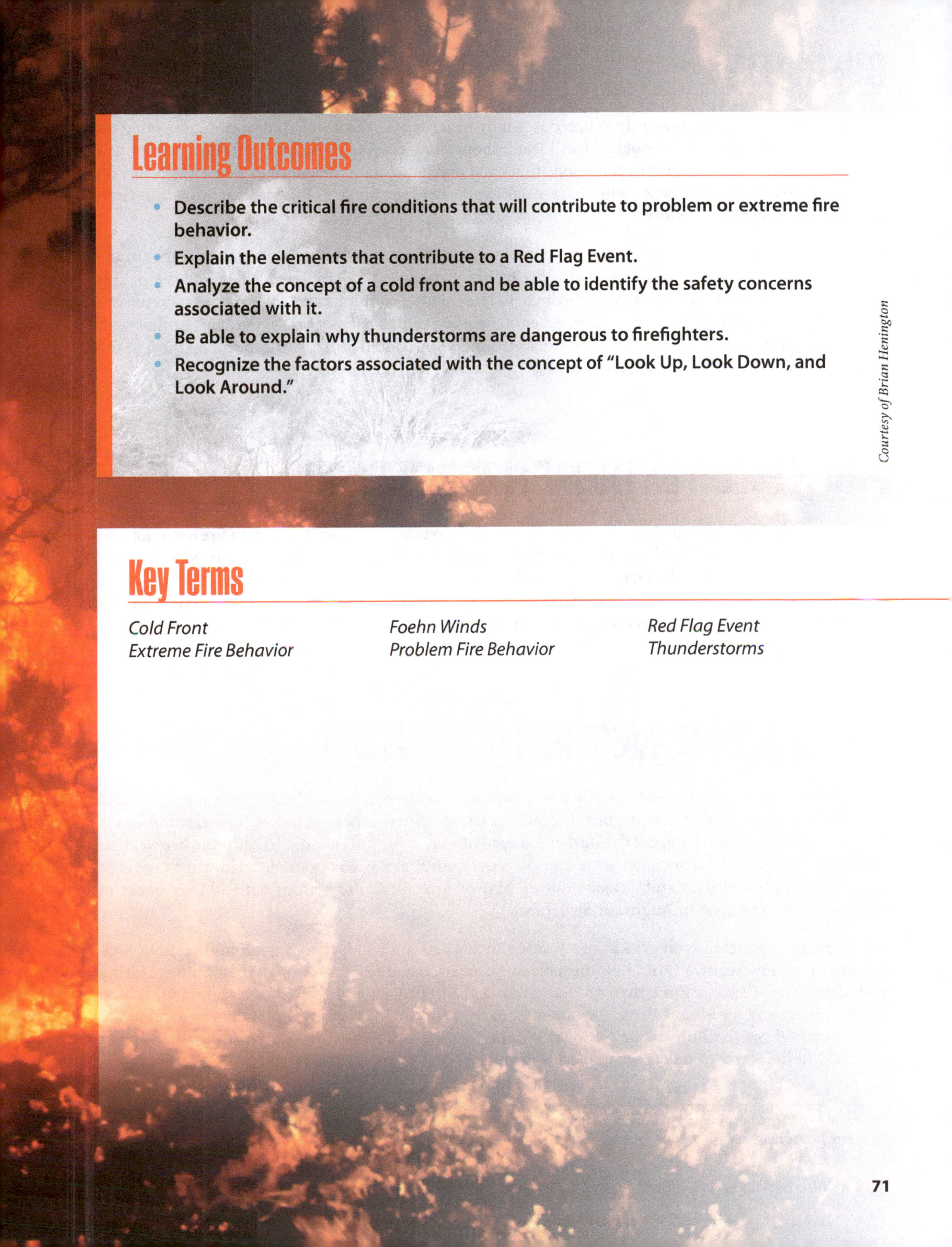

Learning Outcomes

- Describe the critical fire conditions that will contribute to problem or extreme fire behavior.
- Explain the elements that contribute to a Red Flag Event.
- Analyze the concept of a cold front and be able to identify the safety concerns associated with it.
- Be able to explain why thunderstorms are dangerous to firefighters.
- Recognize the factors associated with the concept of "Look Up, Look Down, and Look Around."

Courtesy of Brian Henington

Key Terms

Cold Front
Extreme Fire Behavior
Foehn Winds
Problem Fire Behavior
Red Flag Event
Thunderstorms

OVERVIEW

Wildland firefighting is inherently dangerous. Many hazards and risks will compromise our safety as firefighters. Throughout this book, you will learn about those risks and hazards, but the intent of this chapter is to focus on critical fire conditions that can contribute to a problem or extreme fire behavior. Problem or extreme fire behavior will increase our exposure to the fire itself and present numerous challenges to our suppression efforts.

As a firefighter you are responsible for your personal safety! You must be able to notice changes in the environment, which includes changes in the fire behavior triangle (topography, weather, and fuels). This chapter introduces you to factors that contribute to critical fire conditions and will provide steps for you to identify and monitor those conditions. After completion of this chapter you should be familiar with the concept of "Look Up, Look Down, and Look Around." You should utilize this concept on the first fire you ever fight and continue using it throughout your career as a firefighter.

PROBLEM OR EXTREME FIRE BEHAVIOR

Fire managers classify elevated or increased fire activity into two categories: 1) problem fire behavior and 2) extreme fire behavior. **Problem fire behavior** is described as, "fire activity that presents potential hazard to fireline personnel if the tactics being used are not adjusted. The prediction or anticipation of fire behavior is the key to good tactical decisions and safety."[1] **Extreme fire behavior** is the highest level of burning and presents the highest complexity to firefighter activities. Extreme fire behavior will typically involve heightened heat release rates, intense burning, rapid fire movement, spotting, torching, and/or crowning.

CRITICAL FIRE WEATHER CONDITIONS

Critical fire weather can be best described as conditions the present the highest opportunity for rapid fire growth associated with extreme burning indices. Critical fire weather consists of several components that should be expected to impact the start and spread of a fire at very high rates. In addition, fire weather conditions may also be associated with specific geographical areas. For example, critical fire weather may occur in New Mexico and Arizona during May or June (peak fire season) but will later occur in Washington and Oregon in August or September.

The elements associated with critical fire weather include strong wind, changing wind direction, high temperatures, low relative humidity, unstable atmospheric conditions, and/or dry lightning. During peak days of fire seasons, more than one variable may be in play on any given day. As a firefighter, you should expect any ignition to have the potential to transition into a very fast moving or extreme fire under any of these conditions. The National Weather Service will issue Red Flag Warnings or Watches when any of the above elements are in place and are expected to impact fire behavior.

[1] National Wildfire Coordinating Group, *Introduction to Wildland Fire Behavior, S-190 Student Workbook* (Boise: National Wildfire Coordinating Group, 2006), 3.3.

Red Flag Events

Red Flag Events are an indication that fire behavior may drastically increase or change characteristics. They are of extreme importance to firefighters because of the great danger associated with them. Red Flag Events include:

- **Dry Lightning:** Dry lightning is associated with thunderstorms. Dry lightning events may include some precipitation, but a majority of the threat comes from the potential for new starts because moisture has not reached the ground to subdue new fire starts. Another important concern is the high winds associated with thunderstorm development. (This will be explained in detail later in the chapter).
- **Wind:** Wind indicates events such as cold fronts, foehn winds, and/or thunderstorms that have high winds (15 mph and above) associated with them. In addition, the criteria may also include windy days based on local winds and/or terrain-influenced winds that exceed expected averages.
- **Relative Humidity:** Humidity, especially low humidity, contributes to excessive fire behavior and rapid start and spread of a fire. Low humidity is common with peak fire seasons in specific geographical regions.
- **Fuel Moisture:** Low Fuel Moisture is issued when ten-hour fuels are below 8%.

Red Flag Watch and Warnings

These watches and warnings will be issued by the National Weather Service 12–72 hours prior to a Red Flag Event. The watch may be for specific areas within a geographical area or may include the entire area. The Red Flag Warning is issued when a red flag event is imminent. According the National Weather Service, "the Red Flag Warning is issued immediately when Red Flag conditions are met."[2]

In addition to the elements identified above, there are also specific weather events that have the potential to impact fire behavior and influence the transition to problem or extreme conditions. They include cold fronts, thunderstorms, foehn winds, and/or firewhirls. Firefighters must be aware of these life-threatening weather events and plan their safety measures and tactics accordingly. All of these critical fire weather events have been directly or indirectly involved with firefighter fatalities.

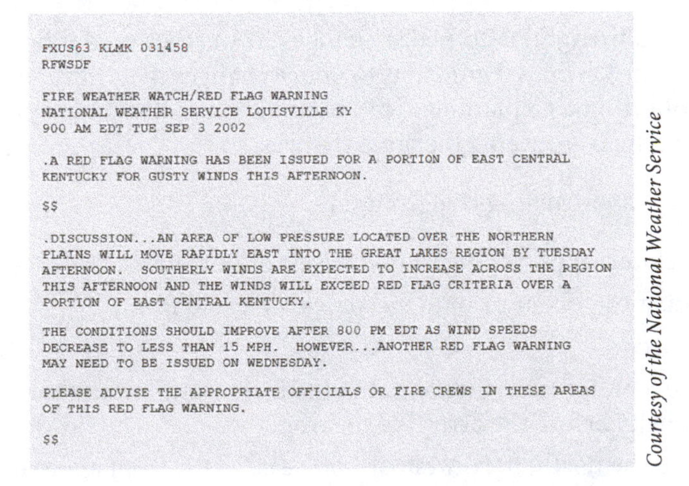

FXUS63 KLMK 031458
RFWSDF

FIRE WEATHER WATCH/RED FLAG WARNING
NATIONAL WEATHER SERVICE LOUISVILLE KY
900 AM EDT TUE SEP 3 2002

.A RED FLAG WARNING HAS BEEN ISSUED FOR A PORTION OF EAST CENTRAL KENTUCKY FOR GUSTY WINDS THIS AFTERNOON.

$$

.DISCUSSION...AN AREA OF LOW PRESSURE LOCATED OVER THE NORTHERN PLAINS WILL MOVE RAPIDLY EAST INTO THE GREAT LAKES REGION BY TUESDAY AFTERNOON. SOUTHERLY WINDS ARE EXPECTED TO INCREASE ACROSS THE REGION THIS AFTERNOON AND THE WINDS WILL EXCEED RED FLAG CRITERIA OVER A PORTION OF EAST CENTRAL KENTUCKY.

THE CONDITIONS SHOULD IMPROVE AFTER 800 PM EDT AS WIND SPEEDS DECREASE TO LESS THAN 15 MPH. HOWEVER...ANOTHER RED FLAG WARNING MAY NEED TO BE ISSUED ON WEDNESDAY.

PLEASE ADVISE THE APPROPRIATE OFFICIALS OR FIRE CREWS IN THESE AREAS OF THIS RED FLAG WARNING.

$$

Courtesy of the National Weather Service

Figure 5-1 *Example of a Red Flag Warning.*

---

1 National Weather Service, *Fire Weather Watches and Red Flag Warnings*, accessed February 3, 2015, http://www.crh.noaa.gov/lmk/product_guide/products/forecast/rfw.php.

Cold Fronts

A cold front is defined "as the transition zone where a cold air mass is replacing a warmer air mass."[3] Cold fronts can occur throughout the year, not only in the winter. Firefighters are concerned with several aspects of these systems: 1) increase in wind speed, 2) shift in wind direction, and 3) unstable atmospheric conditions ahead of the front. Cold front winds played a major role in the firefighter fatalities at the South Canyon Fire in Colorado in 1994. These events are not to be taken lightly! We know how wind influences fire behavior. Now imagine a 180-degree shift in wind direction. How will that impact fire behavior and your safety?

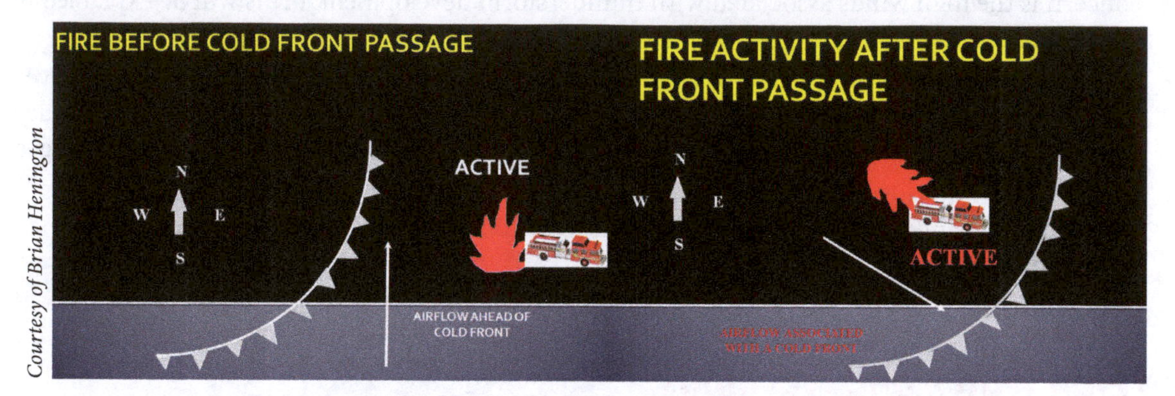

Figure 5-2 *Winds Associated with a Cold Front and the Impact on Firefighter Safety.*

Reasons why cold fronts are dangerous to firefighters:

- Wind direction will abruptly change.
- Strong southerly winds ahead of front will drive the fire head to the north or northeast.
- Winds will shift and change the direction of the head of the fire.
- Rapid drop in relative humidity can occur.
- Winds will be strongest and gusty as the front reaches you.

It is our responsibility as firefighters to notice changes to weather conditions. We do rely on the fire weather forecasts, but your job on the ground is to notice changes that impact fire behavior. Visual indicators do exist with cold fronts. Get familiar with them. If you notice changes, sound the alarm! Don't assume that other firefighters are seeing the same changes.

Some visual indicators of approaching cold fronts are:

- Line of cumulus clouds approaching from the west or northwest.
- Large clouds of dust can occur prior to the arrival of a cold front.
- Dust devils may occur.
- Wind shifts will occur. Strong winds ahead of the front (south, southeast, southwest) will increase in speed and intensity before the arrival of the front.
- Winds will shift to the north, northwest, or northeast as the front approaches.
- Cooler temperatures (after the front) may also be a good indicator for you.

---

[3] University of Illinois, *Definition of a Cold Front*, accessed February 2, 2015, http://ww2010.atmos.uiuc.edu/(Gh)/guides/mtr/af/frnts/cfrnt/def.rxml.

© Taiga/Shutterstock.com

Approaching thunderstorm.

The National Severe Storms Laboratory defines a **thunderstorm** as "a rain shower during which you hear thunder."[4] The National Wildfire Coordinating Group further expands on this definition by explaining, "a storm produced by cumulonimbus cloud and always accompanied by lightning and thunder."[5] A thunderstorm is developed by convective (vertical) life in unstable atmospheric conditions. It produces gusty and very powerful downdraft winds and lightning. Both have been associated with firefighter deaths. It is important to note that not every thunderstorm will have wetting rain that reaches the surface. Those storms are called dry thunderstorms. The threat with these storms is lightning and gusty winds without any moisture to impede new ignitions. As a rule of thumb, thunderstorms are usually short lived with expectancy lasting no more than two to three hours.

Key considerations when a thunderstorm is threatening an emergency incident area:

- Storms will produce lightning. Lightning should not be taken lightly! As a firefighter, we have measures in place that help you reduce the exposure to lightning (Located in the *Incident Response Pocket Guide*—Thunderstorm Safety). In 2014, there were 26 civilian fatalities across the United States related to lightning strikes. The fatalities were most common in Florida (6), Wisconsin (3), Arizona (2), and Colorado (2).[6] Lightning occurs everywhere and it can cause fatalities. Ensure you are familiar with the checklist in the *Incident Response Pocket Guide* and do not continue active suppression activities when thunderstorms are threatening your location.
- Downdraft winds associated with thunderstorms are very volatile and erratic. There have been reports of some downdraft winds reaching 70 mph. The Yarnell Hill Fire (that killed 19 of our finest firefighters from the Granite Mountain Hotshots) was highly influenced by downdraft winds from a thunderstorm. The winds influenced fire spread, intensity, and direction of the fire head.
- Fires may produce their own thunderstorms. Fires burning under extreme conditions can create their own thunderstorm from the convection column produced by the fire.

[4] National Oceanic and Atmospheric Administration: National Severe Storms Laboratory, *Severe Weather 101: Thunderstorm Basics*, accessed February 6, 2015, http://www.nssl.noaa.gov/education/svrwx101/thunderstorms/.

[5] National Wildfire Coordinating Group, *Wildland Fire Behavior*, 2C.26.

[6] National Weather Service, *Lightning Safety*, accessed February 12, 2015, http://www.lightningsafety.noaa.gov/fatalities.htm.

Like cold fronts, thunderstorms will have visual indicators that allow firefighters to witness the approaching storm. Pay attention and make sure other firefighters are aware when you notice any of the following:

- Tall, building cumulus clouds that have a look similar to cauliflower.
- Thunderstorms will have a dark flat base.
- Clouds that have virga (rain that does not reach the ground) or rain falling from the bottom of cloud.
- Ice crystal top usually in anvil shape with a fuzzy appearance.
- Look for the anvil top of the thunderstorm. This will allow you to determine the direction the storm is moving.

Thunderstorm

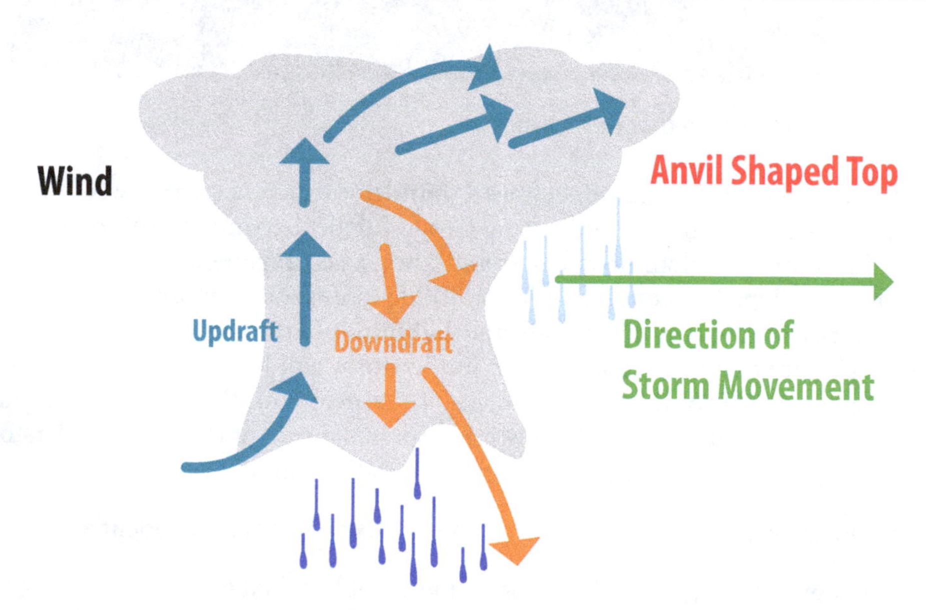

Figure 5-3 *Anvil Top of a Thunderstorm.*

Santa Ana winds influencing fire activity in California.

Foehn winds are dry, strong winds that are common to specific geographical areas across the United States. Foehn winds occur over high mountain ranges and flow down the lee side of those mountains. They are not short lived like a thunderstorm. Some foehn winds can occur over several consecutive days or weeks. They have been reported to reach speeds up to 90 mph. Common wind speeds average between 40 to 50 mph. Hot, dry, and very volatile winds are a recipe for disaster when a fire is introduced into the picture. The list below identifies the most common foehn winds found in the United States.

Name of Foehn Wind	Geographical Area or Location
Santa Ana and Sundowner	Southern California
Chinook wind	East side of the Rockies and east side of the Sierra Nevadas
Wasatch wind	West side of Wasatch Range in Utah
Mono or North	Central and Northern California
East wind	Western Oregon and Western Washington

Firewhirls

Firewhirls are not going to occur on every fire you fight. However, you should be aware that they can occur under extreme conditions. I was once asked how to explain a firewhirl. I thought about it and decided to use this example. It's a small fire tornado. For extra emphasis and to wake students up, I say, A TORNADO WITH FIRE IN IT! I know I am over exaggerating the concept, but the point is that firewhirls are very dangerous.

© Tom Reichner/Shutterstock.com

The early stages of a firewhirl. Carlton Complex—Washington—2014.

Firewhirls occur under extreme burning conditions. Their energy is derived from the surface fire. Once they lose the energy of the surface fire, then, in theory, they will lose their power. Firewhirls are dangerous because they can move in any direction and they are a major source for transporting fire brands, which will create spot fires. Firewhirls will also have increased and gusty winds associated with them.

Dust Devils

Dust devils present the same challenge as firewhirls except they do not have active fire associated with them. As you have learned, dust devils are an indication of unstable atmospheric conditions. Dust devils can move through a fire area and cause major safety concerns. Just like firewhirls, they can transport fire brands. The wind associated with them is intense. Fire behavior will drastically increase. Dust devils will occur on hot, dry days. If you notice the development of dust devils throughout the day, you should prepare for increased and elevated fire activity.

YOU MUST MONITOR FIRE BEHAVIOR

It is the responsibility of every firefighter (not just the fire managers) to monitor fire behavior. We must base our actions on the current and expected behavior of the fire. We must further evaluate topography, fuels, and weather conditions to ensure that the best solution is in place for our tactical activities. Fire does not behave as it did in the movie *Bambi*. Lightning does not strike a tree and immediately cause a full-blown crown fire. A process has to occur before a crown fire can or will occur. We must notice and inform others of the indicators that contribute to extreme fire behavior.

Extreme fire behavior will typically involve heightened heat release rates, intense burning, rapid-fire movement, spotting, torching, and/or crowning. If you are involved in a fire where any of these events are occurring, than you need to back up and regroup. Constantly evaluate your surroundings and fire behavior. Notice the changes that are occurring and make them known. Ensure the most appropriate and safest tactical solution is in place. As an entry-level firefighter, you will not be making the tactical decisions used on a fire; however, you are responsible for your own safety. So if you notice any of these extreme fire behavior indicators, ensure others also notice them. Call it out and don't assume others are seeing what you are seeing. If we all use this approach, then we can ensure that all fighting the fire are aware and informed.

Courtesy of Brian Henington

Firefighter sounding the alarm on increased fire behavior.

LOOK UP, LOOK DOWN, AND LOOK AROUND

The concept of "Look Up, Look Down, and Look Around" was created by the National Wildfire Coordinating Group to provide firefighters with the basic skills and knowledge necessary to identify factors that may contribute to elevated fire behavior. The title of this concept illustrates how critical it is for firefighters to evaluate the entire fire picture. It is easy to become distracted with our firefighting efforts. This concept demands that we evaluate our surroundings at all times. Take a tactical pause, and Look Up, Look Down, and Look Around. If we all do this, we will be able to plan our actions accordingly based on the current and expected behavior of the fire. The concept is explained in detail below.

1. **Fuel Characteristics:** Fuel drives fire. Expect extreme fire behavior when the following condition or conditions exist.

Concern/Condition	Explanation
Continuous fine fuels © kenjii/Shutterstock.com	• Expect increased and fast moving fires. • When wind and slope are introduced with fine fuels, expect greater impact on spread, intensity, and speed.

(continued)

Concern/Condition	Explanation
Heavy Loading of Dead and Down Fuel  © basel101658/Shutterstock.com	• Larger fuel, under dry conditions, will have great impact on intensity and the heat release rate of the fire. • The laddering effect. Abundant ladder fuels will allow fire to transition from the surface to the crowns.
Ladder Fuels  © Alin Brotea/Shutterstock.com	• The laddering effect. Abundant ladder fuels will allow fire to transition from the surface to the crowns.
Tight Crown Spacing  Courtesy of Brian Henington	• The crowns of trees or brush are intertwined or are very close to each other. This will support crown fires.

(continued)

Concern/Condition	Explanation
Overabundance of snags will influence fire behavior.  Courtesy of Brian Henington Insect-infested trees. Notice the standing red slash on the trees.  Courtesy of Brian Henington	• Firebrand sources—fires that burn certain fuel species may result in more spot fires. Examples include: Pine bark plates, oak leaves, and manzanita leaves. • Snags—Dead or partially dead standing trees are more apt to burn. Snags are considered extremely dangerous. • Frost kill—early spring frost kills leaves. Leaves are more willing to burn. • Bug kill—dominated areas across the west. Dead fuel available for consumption. • Preheated Canopy—surface fire (through convective and radiant heat) have dried fuels making them more susceptible to burn. • Unusual fine fuels—fire spread and intensity can be extreme. • High dead level of dead fuels--More dead fuels in the fire area, the more potential for intense fires.

2. **Fuel Moisture:** The amount of moisture in any given fuel will have a direct impact on the availability of that fuel to ignite and sustain combustion.

Concern/Condition	Explanation
Low Relative Humidity—below 25%	Fine fuels are highly influenced by relative humidity. The lower the relative humidity, the more available the fuels are to burn. Low relative humidity is an indication of a Red Flag Event.
Low 10 Hour Fuel Moisture Content. Courtesy of Brian Henington	Ten-hour fuels are a good indication of fire behavior activity. Low moisture content in these fuels will increase fire behavior and add to the intensity of the fire.

(continued)

Concern/Condition	Explanation
Drought Conditions.  Courtesy of Brian Henington	Fuels are weakened. The lack of moisture decreases the fuel moisture in live fuels. These fuels can and will support rapid fire growth.
Seasonal Drying.  Courtesy of Brian Henington	As the days get hotter and drier, expect elevated fire risk due to the lack of moisture in available fuels.

3. **Fuel Temperature:** The higher the temperature of individual fuel, the more apt they are to burn and sustain combustion.

Concern/Condition	Explanation
High Temperatures (above 85^0 F)	Fuels are more willing and able to burn under high temperatures. Expect rapid fire development with prolonged days of 85^0 F and above.
Fuel Exposed to Direct Sunlight. Courtesy of Brian Henington	Decreases fuel moisture which allows fuels to sustain combustion.

(continued)

Concern/Condition	Explanation
Aspect Influences.  Courtesy of Brian Henington *South aspect. Fuels are prime and ready to burn.*	Will have direct impact of fuel temperatures. South and southwest aspects have most direct sunlight throughout the day. Expect elevated fire behavior on these slopes.

4. **Topography:** You should fully understand that topography influences play a major role in fire behavior. Always be aware of your surroundings and your location as it relates to topographical features.

Concern/Condition	Explanation
Steep Slopes (over 50%)  Courtesy of Brian Henington *If a fire was introduced on this steep slope, activity would be increased and elevated.*	Expect rapid fire spread. Convective lift plays a critical role on steep slopes. The opportunity for rolling debris is drastically increased. Finally, the wear and tear on a firefighter's body is also increased.
Chimneys or Chutes 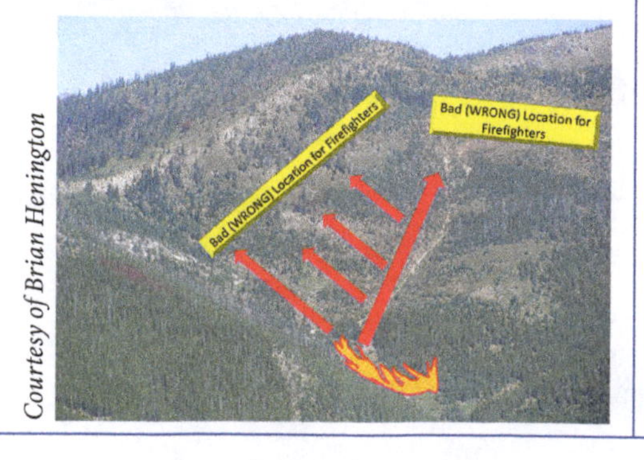 Courtesy of Brian Henington	Funnels fire. Chimney effect is in play. Expect rapid rates of spread. Avoid them!

(continued)

Concern/Condition	Explanation
Saddles	Funnels fires. Avoid tactical activities in a saddle.
Box Canyons	DON'T go in them! Funnels fires using the chimney effect. Wind is increased in these canyons. Fire behavior will be elevated.
Narrow Canyons	Greater impact of radiant and convective heat to impact adjacent (not burning) slope. Spotting is increased, which will increase the ignition opportunities to develop combustion. Fires on both slopes will burn toward each other making the bottom of the canyon a death trap.

5. **Wind:** Wind has immediate impact on fire behavior. Firefighters have lost their lives due to wind influences impacting fire behavior. Always monitor wind conditions to include speed and direction. As a firefighter, you must ensure that other firefighters are aware of any of these changes identified below.

Concern/Condition	Explanation
Surface Winds (above 10 mph).  *Courtesy of Brian Henington*	Winds over 10 mph will influence fire spread and direction. Expect elevated fire behavior because of increased oxygen available for the fire. These winds are strong enough to transport fire brands.
Lenticular Cloud Development.  *Courtesy of Brian Henington*	Good indication of a windy day. These clouds usually form over high mountain regions. They have been described as "looking like a spaceship." If you notice them in the morning, you should expect high winds aloft later in the day to potentially impact the surface, thus impacting your fire.
Approaching Cold Fronts	Expect a wind shift and increase in wind speed as the front approaches your fire.

(continued)

Concern/Condition	Explanation
Thunderstorm Development © Dario Lo Presti/Shutterstock.com	Expect strong and gusty downdraft winds. Expect lightning.
Battling or Shifting Winds	Expect a change in wind speed and direction. Fire behavior will change and/or increase.
Sudden Calm	Indication that a change is coming. Have you ever heard of the saying, "Calm before the storm?" A wind change is coming.

6. **Atmospheric Stability:** Every firefighter should be aware of the atmospheric conditions during fire activities. An unstable atmosphere will support large fire development and elevated fire activity.

Concern/Condition	Explanation
Unstable Atmosphere Courtesy of Brian Henington *Two visual indicators are occurring in this photograph. 1) Inversion (stable atmosphere) is lifting. 2) Smoke column rises vertically as the change from stable to unstable is occurring.*	Fires burn with the most intensity under unstable atmospheric conditions. Pay attention to visual indicators of an unstable atmosphere. Be prepared for active burning. • Smoke rises vertically • Good visibility • Gusty winds • Dust devils • Cumulus clouds • Castellatus clouds in the morning • Inversion lifting

(continued)

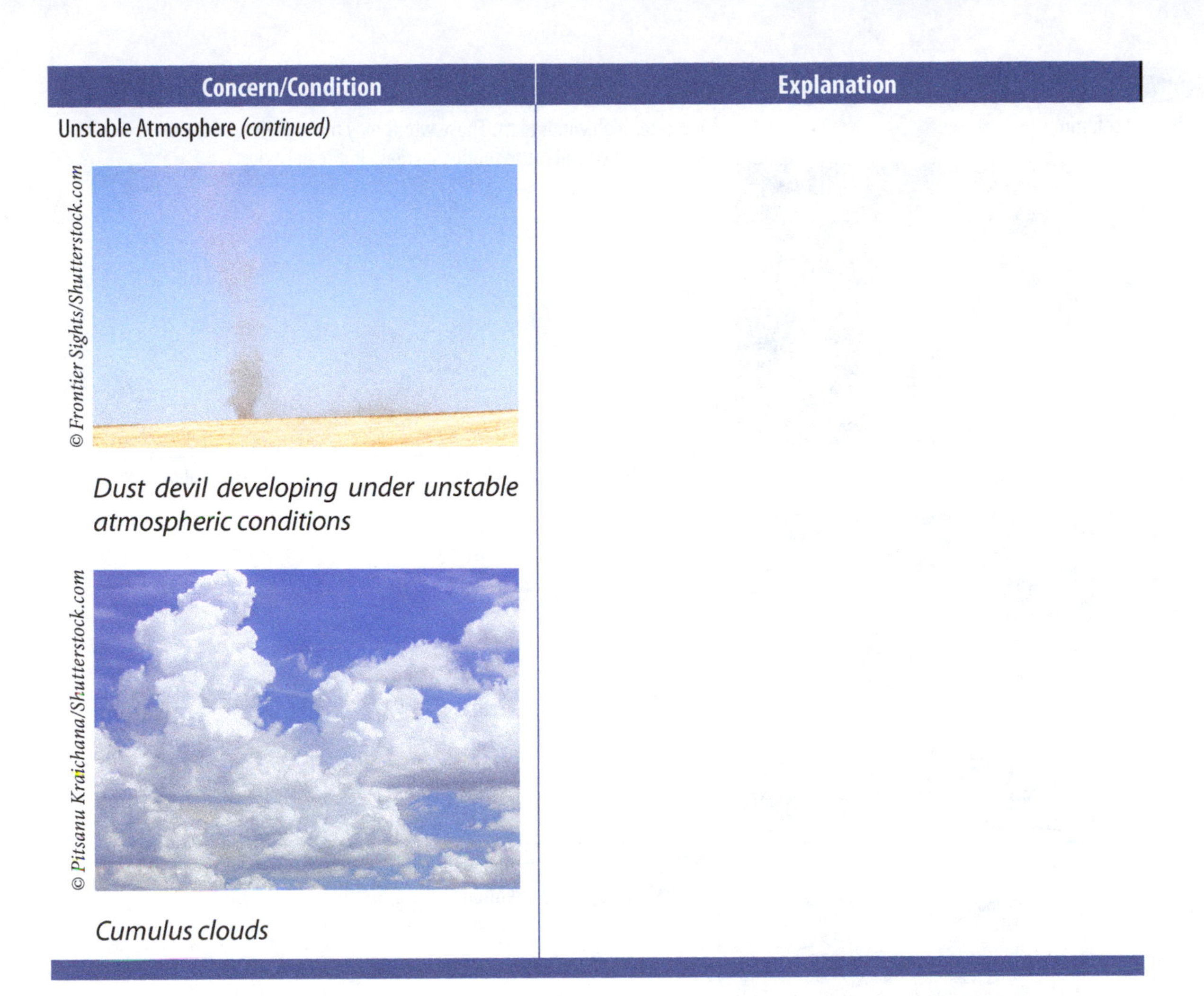

Concern/Condition	Explanation
Unstable Atmosphere *(continued)* © Frontier Sights/Shutterstock.com *Dust devil developing under unstable atmospheric conditions* © Pitsanu Kraichana/Shutterstock.com *Cumulus clouds*	

7. **Fire Behavior:** Fire behavior is an immediate visual indicator of elevated or increased fire behavior. Every firefighter must monitor and evaluate fire behavior to ensure risks can be identified and properly mitigated. If you see something concerning, Sound the Alarm! Follow the chain of command and ensure others are aware of what you are seeing.

Concern/Condition	Explanation
Leaning column. Courtesy of Brian Henington	Indicates wind driving or influencing the fire.

(continued)

Concern/Condition	Explanation
Sheared column  © James Mattil/Shutterstock.com	Indicates high winds aloft. Those winds may make their way to the surface, which will cause sudden increase in fire behavior.
Well-developed column.  Courtesy of Brian Henington	Extreme burning. Pay particular attention to how quickly the column develops.
Changing column  Courtesy of Brian Henington *Smoke column transitions from white to dark gray, then black. Dense smoke means fire behavior is increasing.*	Color and volume of column will indicate intensity and fuels involved. Ensure you pay close attention to the column.

(continued)

Concern/Condition	Explanation
Trees torching.  *Courtesy of Brian Henington*	Fire activity is increasing. Torching has to occur before a crown fire. Ensure crew members are aware that torching is occurring. Reevaluate LCES (discussed in later chapters).
Smoldering fires picking up.  *Courtesy of Brian Henington*	Fire activity is increasing. Again, ensure crew members and other fire personnel are aware of what you are seeing.
Small fire whirls begin	Fire activity and intensity is increasing. Ensure crew members and other fire personnel are aware of what you are seeing.
Frequent spot fires	Indication of extreme fire behavior. Numerous spot fires may cause control issues and jeopardize firefighter safety.

SUMMARY

Problem and extreme fire behavior present the greatest challenges to firefighters. Extreme fire behavior is the worst-case scenario. Every fire has the potential to elevate, increase, and transition to extreme. Your responsibility is to ensure you are noticing the changes and events that will lead to extreme burning. If you follow and use the concepts in Look Up, Look Down, and Look Around you will ensure that your safety and the safety of fellow firefighters is the top priority. Learn to use your *Incident Response Pocket Guide* to help you evaluate and monitor conditions related to fire activity. You are expected to base all your actions on current and expected behavior of the fire. Using the process identified in this chapter will ensure that you are adhering to this critical rule of wildland firefighting.

KNOWLEDGE ASSESSMENT

1. What are the four Red Flag Events that can lead to a Red Flag Warning or Watch?
2. What agency is responsible for issuing Red Flag Warnings or Watches?
3. What is a cold front? What are the dangers associated with cold fronts? What major fire fatality occurred during an approaching cold front?
4. Thunderstorms are very dangerous to firefighters. What are the two major concerns identified in this chapter?
5. What historical firefighter fatality occurred due to an approaching thunderstorm?
6. List the common types of foehn winds that occur in the United States.
7. Why are firewhirls dangerous?
8. List the seven Look Up, Look Down, Look Around factors that must be monitored by wildland firefighters.

EXERCISES

1. Using your *Incident Response Pocket Guide* (IRPG), find thunderstorm safety. What page of the IRPG did you locate this safety checklist? List for your instructor the steps involved in this checklist.
2. Using the National Weather Service as a resource, locate a lightning related fatality that occurred in your area. How did it occur? What activity was the individual(s) doing when they were struck?
3. Research the National Fire Center's *Historical Wildland Firefighter Fatality Reports*. Search the database by type of accident. How many firefighters died in the line of duty due to lightning strikes? What state(s) did they occur in?
4. Conduct an Internet search on firewhirls. Provide your instructor with the website. You do not have to write a discussion on them. The question we have for you is: What did you think about this phenomenon?

BIBLIOGRAPHY

National Wildfire Coordinating Group. *Glossary of Wildland Fire Terminology PMS 205*. Boise: National Wildfire Coordinating Group, 2014.

National Wildfire Coordinating Group. *Introduction to Wildland Fire Behavior S-190*. Boise: National Wildfire Coordinating Group, 2006.

National Weather Service. *Fire Weather Watches and Red Flag Warnings*. http://www.crh.noaa.gov/lmk/product_guide/products/forecast/rfw.php.

National Weather Service. *Lightning Safety*. National Weather Service. http://www.lightningsafety.noaa.gov/fatalities.htm.

National Oceanic and Atmospheric Administration: National Severe Storms Laboratory, *Severe Weather 101: Thunderstorm Basics*. http://www.nssl.noaa.gov/education/svrwx101/thunderstorms/.

University of Illinois. *Definition of a Cold Front*. http://ww2010.atmos.uiuc.edu/(Gh)/guides/mtr/af/frnts/cfrnt/def.rxml.

CHAPTER 6

FIREFIGHTER READINESS AND CORE ESSENTIALS

Learning Outcomes

- Identify and list the core personal protective equipment needed to participate in wildland fire suppression activities.
- Explain the personal gear needed for extended or lengthy fire assignments.
- State the importance of professionalism and how it relates to firefighters.
- Explain the process of conflict resolution and how the process should be implemented among firefighters.
- Recognize the significance of teamwork and how it supports safe and efficient firefighting operations.
- Develop a physical fitness regime and explain why fitness and health play a critical role in a firefighter's life.

Courtesy of Steven Ikeda

Key Terms

Accountability
Agency Affiliation
Conflict Resolution
Cultural Differences
Ethics
Fatigue
Fireline Fitness
Flame Resistant Clothing
Nomex™
Pack Test
Personal Protective Equipment
Professionalism
Teamwork

OVERVIEW

Personal protective equipment is considered mandatory protection equipment or safety gear that must be worn during all fire suppression activities. The purpose of this equipment or gear is to ensure the firefighter's personal well-being is prioritized and that safety precautions are in place to mitigate exposure to hazards. It is your responsibility to ensure the equipment or gear assigned to you is maintained in the most effective manner possible. All firefighters must be ready to respond to an incident within a moment's notice. It's not the agency or fire departments responsibility to ensure you are ready for a dispatch. Firefighters should have their gear and personal protective equipment prepped, maintained, and ready for dispatch at all times.

Courtesy of Brian Henington

Wildland Firefighter: Steven Ikeda.

Firefighting also requires a set of essential elements that need to be maintained by all firefighters. Firefighting is an honorable profession. The public holds us in high accord; therefore, we must ensure we behave and conduct ourselves in the most professional manner possible. Our appearance and behavior is a reflection on the agency we work or volunteer for. Maintaining a high standard of expectations, behavior, and attitude is expected by our employers. This chapter will also explore key essentials of firefighter behavior, attitude, and appearance as it relates to our employment.

Finally, any form of firefighting requires the firefighter to be in top physical conditioning. Just as an athlete prepares and develops their bodies for competition, we must also prioritize the importance of a rigorous physical fitness regime to ensure we can perform at the highest level. We must also maintain a healthy diet to ensure our bodies can recover from the arduous activities we perform. This chapter will introduce you to physical fitness concepts and techniques needed to make you an efficient, safe, and effective firefighter.

PERSONAL PROTECTIVE EQUIPMENT

Personal protective equipment is described as, "equipment worn to minimize exposure to serious workplace injuries and illnesses."[1] Personal protective equipment is often referred to as PPE. The hazards and exposure to risks are considered high in wildland firefighting. PPE has been designed to reduce direct exposure to elements that may directly or indirectly harm you. The harm can be immediate or can occur over years of exposure. As a firefighter, you need to ensure that all of the required PPE is worn at all times during the suppression activities of a fire.

Established wildland firefighting standards require several mandatory PPE items that must be worn during all fire suppression activities. (Note: some items are only intended for specific activities. They will be identified below). For the purpose of this chapter, we have identified the core PPE you will need to

[1] United States Department of Labor: Occupational Safety & Health Administration, *Definition of Personal Protective Equipment*, accessed on February 8, 2015, https://www.osha.gov/SLTC/personalprotectiveequipment/.

participate in fire suppression activities. If a PPE item is mandatory by established standards or codes, then the agency, fire department, or company you work for will provide these elements (with some exceptions—mainly your fire boots).

In addition, all clothing and specific equipment worn by firefighters on a wildland fire must adhere to standards established by the National Fire Protection Association (NFPA). The minimum standards are established in *NFPA 1977: Standard on Protective Clothing and Equipment of Wildland Firefighting* (current edition 2011 with next edition occurring in 2016). The purpose of this standard is best explained by NFPA:

> *Protective clothing and equipment covered include garments, helmets, gloves, footwear, and goggles, chainsaw protectors, and load-carrying equipment. Provisions apply to certification, product labeling, instructions for the user, design criteria, performance requirements, and testing methods for everything from resistance to tearing, flame, conductive heat, puncture, and corrosion to tests for dexterity, grip, burst strength, and abrasion.*[2]

The most important concept we can take from the NFPA is that you cannot wear any gear you want on the fireline. If the items are mandatory PPE or required to protect from exposure, then they must meet these NFPA standards. You cannot wear an outer jacket on a fire unless that jacket meets the standards.

Figure 6-1 *PPE for the Wildland Firefighter.*

Mandatory Personal Protective Equipment (PPE) All of these items should be worn at all times (unless otherwise noted).

You should maintain all of your assigned gear in the best possible manner and maintain it using the instructions provided with the gear. Remember, this equipment typically belongs to an agency or fire department. We need to make sure it is maintained in the best possible condition.

1. **Hard Hat**: Protects your head from falling or rolling debris. Must be NFPA approved. Hard plastic only. Metal hard hats are not allowed because of the concern with lightning and electricity. The two common styles of hard hats are a full brim or cap (baseball cap). Some firefighters prefer

[2] National Fire Protection Association, NFPA 1977, *Standard on Protective Clothing and Equipment for Wildland Firefighting: 2011 Edition.*

the full brim to help with sunburn issues; whereas others refer the cap style (do not wear backwards). The color of the hard hat typically identifies the agency, crew, or fire department you work for. Note: Do not add stickers, decals, or write on the helmet without permission from your fire department, agency, or organization.

2. **Flame Resistant Clothing**: **Nomex™** or **Aramid.**

   Flame Resistant Clothing: Definition—resistant to burning, but is not considered 100% flame-proof. Flame resistant clothing was created to reduce exposure from convective and radiant heat and direct flame contact. Extended exposure to any of these elements will not guarantee protection. There have been documented events where flame resistant clothing has burned or melted to firefighters.

   Nomex™: Synthetic fire resistant clothing. Nomex™ is the trade name. The generic version of Nomex™ is Aramid.

   Wildland fire gear is made to reduce physical wear and tear on firefighters. Fire pants will have more room (for climbing hills and steeping over logs) than a normal off duty pant. Ensure your fire shirt and/or pants do not have oil, gas or other combustible materials on them. Oil or gas on clothing is very common for chainsaw operators. Any significant amount of oil or gas on your fire resistant clothing will make them ineffective and could make them combustible. There is a high potential to be removed by a Safety Officer from the fireline if you have oil or gas on your clothes.

   a. Shirt: Firefighting shirts must be long sleeve (do not roll up your sleeves or unbutton your top button (this can expose your undershirt or chest). Keep shirt buttoned. The most common color for fire shirts is yellow; however, it is not uncommon to see khaki, red, or blue fire shirts. The yellow fire shirt provides a visual indicator of your location. Highly-visible shirts are recommended. Inmate crews across the United States will wear red or orange fire clothing to identify them as inmate crews.

   b. Trousers (pants): There are numerous brands and types of fire pants available. The most common style today is the BDU (Battle Dress Uniform) style with multiple pockets available to carry gear or supplies.

3. **Eye protection**: Must be shatter-resistance and OSHA approved. Eyes are extremely vulnerable to blowing dust, debris, and wood chips from chainsaws or chippers. Safety glasses must be worn during both day and night operations. Clear or yellow tinted glasses are recommended for night-time operations. You should have both clear and dark glasses in your fire pack or kit.

   Prescription glasses must be shatter resistant and OSHA approved. You are responsible for your own prescription glasses. It's a good idea to have an extra set of prescription glasses in your kit in case the original ones break. Contact lenses are not considered eye protection and may become problem because of the dusty and smoky environment we work in. Contact lenses are not recommended on the fireline; instead, use prescription glasses.

HELPFUL HINT

Allow the agency you work for supply you with eye protection. Wearing expensive sunglasses or prescription glasses on the fireline is not recommended. If they become scratched or damaged, you will have to replace them. If this happens to agency eye protection, the agency will replace them.

4. **Hearing protection**: Must be worn when working with or near loud equipment. Some firefighters prefer earplugs instead of ear muffs. The choice is yours. Have extra earplugs in your fire pack. When working with loud equipment, try wearing earplugs with ear muffs. Examples of loud and very common activities: chainsaws, aircraft, dozers, or portable pumps.

PROTECTING YOUR HEARING IS ESSENTIAL!

I have many friends who have lost some of their hearing due to the loud environment we work in. Most activities on a fire are loud. Always protect your hearing!

5. **Gloves**: Keep an extra set of leather gloves with your fire shelter and in your fire pack. Many injuries and burns have occurred because firefighters were not wearing their gloves. Calluses and blisters are a very common occurrence to firefighters. Gloves will help protect from these issues.
6. **Fire Boots:** Must be a minimum of 8" tall. Made from 100% leather with **NO** steel toes. The boots must also include a deep tread that provides for traction. The boots should also have leather shoelaces (always carry extra bootlaces in your fire pack). Most agencies will not purchase your fire boots. There are numerous manufactures of fire boots. Talk to other firefighters for their recommendations. Ensure the boots you purchase meet the minimum wildland fire requirements. Synthetic boots (like military boots), steel toes, and zippered boots are not allowed on the fireline.

AUTHOR RECOMMENDATION

Take care of your feet! The boots will be a major expense to you. Make sure you try them on and you like the feel and comfort. You will be in your boots up to 16 hours a day. In addition, also learn how to properly maintain and care for your boots. Again, a mentor or seasoned firefighter can show you the proper techniques for caring for your boots.

Courtesy of Brian Henington

WILDLAND FIRE BOOT REQUIREMENTS

1. 100% leather
2. Minimum of 8" tall
3. No steel toe
4. Deep-tread sole for traction
5. Leather shoe laces

7. **Fire Shelter**: New Generation Fire Shelter. (Fire Shelters will be discussed in detail in later chapters). Keep an extra set of gloves with your shelter.

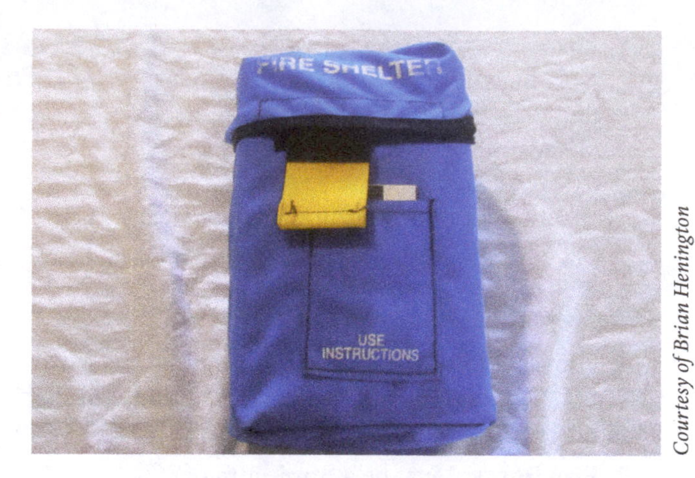

Courtesy of Brian Henington

New Generation Fire Shelter in Protective Case.

8. **Canteen or camelback**: Stay hydrated! Dehydration is a major issue to firefighters. Ensure you have adequate water available for a 16-hour shift. Note: Most new fire packs have slots for a camelback, which make life a little easier. Standard issue canteens are plastic. The National Wildfire Coordinating Group recommends that you drink a minimum of one gallon of water per day.
9. **Chinstrap**: Usually comes with the hard hat. You must have a chinstrap for certain activities such as approaching and riding on a helicopter, deployment of a fire shelter, or water or retardant drops threatening your location. A majority of firefighters do not wear a chinstrap unless the situation requires it. It is stored on the helmet over the brim of the hardhat.

Courtesy of Brian Henington

10. **Fire Pack or Harness to Carry Fire Shelter**: Your fire pack is an essential tool and carries your fire shelter. Your fire shelter must be accessible and you should have it available at all times. Similar to boots, there are numerous manufactures of fire packs. Your fire pack is usually provided by the agency you work for. Make sure it fits properly before use. A comfortable fire pack makes life easier on the firefighter. Engine crews sometimes wear web packs that do not include a large pouch for gear. If you elect to wear web gear, you need to ensure your fire shelter is easily accessible.
11. **Headlamp**: New headlamps are bright and small in nature. Ensure they are in your fire pack at all times. In addition, make sure you have extra batteries.

Author Note: I was on a fire where one firefighter did not have a headlamp. We were only supposed to work during the day; however, as the saying goes, "Don't count on it." We did not disengage the fire until 2 am. As we were walking down hill to our camp, the firefighter fell on a boulder, breaking his leg. He claimed he never saw the boulder. The lesson: always have a headlamp available and make sure it works.

12. **First Aid Kit**: Ensure your kit is fresh and up-to-date with all the required supplies. The kit is designed for minor injuries.
13. **Incident Response Pocket Guide:** The most important reference material we can use, it is the guide to the concepts and safety measures you may forget.

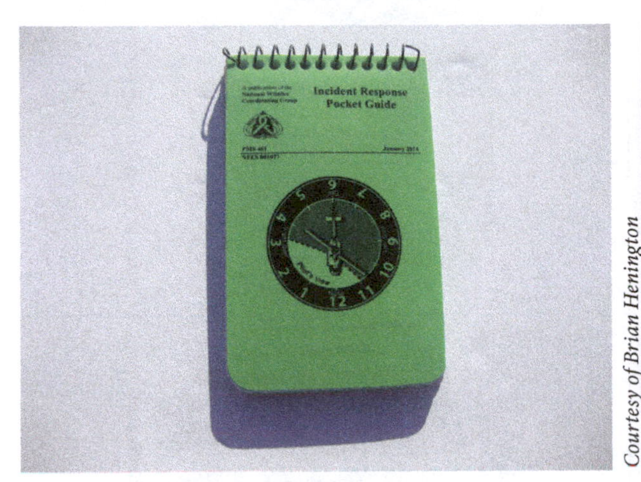

Courtesy of Brian Henington

Incident Response Pocket Guide (IRPG).

Courtesy of Brian Henington

Firefighter Mark Naranjo protecting his airway in smoky conditions.

Additional Items (for your fire pack or overnight bag)

The following additional items should be part of your kit; however some of the items listed below are required PPE for certain activities: Appendix A: Recommended Gear for a Wildland Fire Assignment (Checklist) has a list of recommended items firefighters should carry in their overnight bags for long duration fire incidents.

Chainsaw Chaps: Only required if you are operating a chainsaw and/or supporting chainsaw operations. Make sure the chaps are the correct length. They should cover the entire area of your leg from your waist to your boots. Short chaps (only covering to your shin) are not effective and defeat the purpose of the chaps.	Sawyer Brian Filip.  Courtesy of Brian Henington

(continued)

Food: Carry food in your fire pack. Examples include Meal: Ready to Eat (MRE's), protein bars, or small canned food.	Meal: Ready to Eat.  *Courtesy of Brian Henington*
Flat File: Used to sharpen your hand tools. Minimum of 10".	Flat File—12".  *Courtesy of Brian Henington*
Bandanna: Used by most firefighters to protect airways. It is also used on your head to ensure the hard hat stays tight and to keep sweat out of your eyes. Can also be useful if you run out of toilet paper.	 *© nito/Shutterstock.com*
Socks: Take care of your feet! Just like boots, socks will be your expense. There are numerous types of socks available. Seasoned firefighters will have good ideas for the best socks. Author recommendation: Get socks that fit your foot snug and that are tall. Blisters are a major issue. Tight socks may help eliminate the friction that causes blisters. Invest in good socks that help wick moisture away from your feet. Your feet will take considerable wear and tear and pounding. Learn how to protect them to eliminate the wear and tear. Try blister cream. Whatever you can do to protect your feet and make your fire assignment as pleasurable is the key here. Most firefighters complain about either blisters or soreness to their feet.	Tall, snug socks are recommended.  *Courtesy of Brian Henington*

(continued)

Overnight or Personal Gear Bag: Used to store your additional clothes and gear. Bag should be big enough to hold 14 days' worth of gear. Usually provided by your agency or fire department.	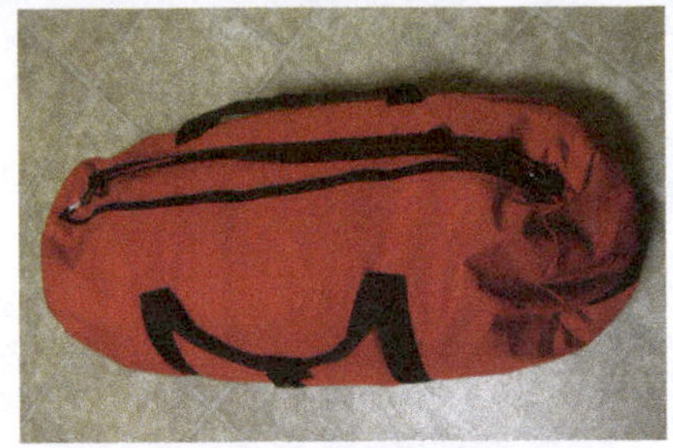 *Courtesy of Brian Henington*
Hard Hat Shroud: Some firefighters prefer to use the shroud to protect from flying embers, radiant heat, and sunburns.	Neck shroud.  *Courtesy of Brian Henington*
Signal Mirror: Key tool for working as a lookout or working with aircraft. Mirrors are ideal for signaling your location. Ensure your mirror is in a protective case.	Standard signal mirror. 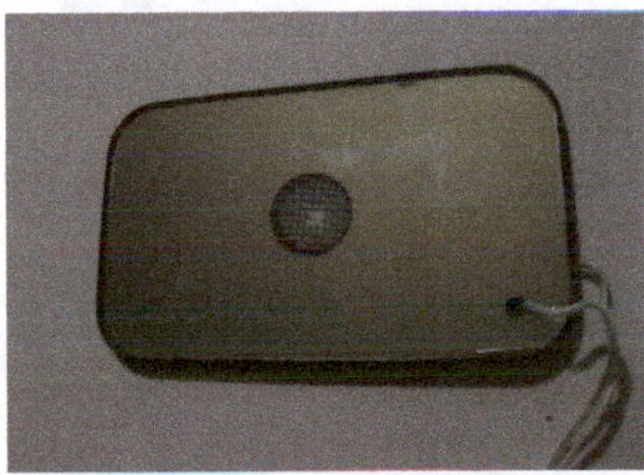 *Courtesy of Brian Henington*
Compass: Used to determine location, bearings, and wind direction.	 *Courtesy of Brian Henington*

(continued)

GPS Unit: Common and useful tool to identify locations.	GPS Unit.  © Jeff Metzger/Shutterstock.com
Hand Held, Field Programmable Radio: Not all firefighters will be issued radios due to the cost. Supervisors will have radios. Radios should be field programmable (explained in later chapters).	Handheld radio.  Courtesy of Brian Henington
Radio Chest Pack: Carry radios across chest for easy use and access and to protect from wear and tear.	Radio chest pack with radio.  Courtesy of Brian Henington
Fire Jacket or Coat: Approved NFPA jacket. Great for prescribed fires and night operations.	Nomex™ Fleece NFPA approved jacket.  Courtesy of Brian Henington

(continued)

Sweatshirt: Used during non-fire activities. Good for cool mornings. Individual crews will typically have their own sweatshirts. Also great for crew identification and builds pride among crews.	Hooded sweatshirt. 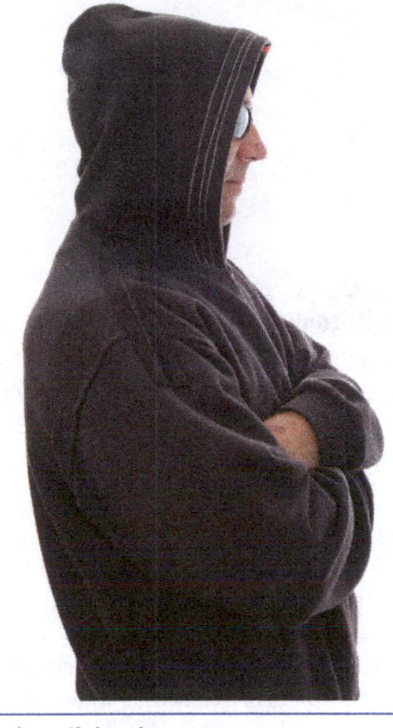 © dabjola /Shutterstock.com
Hat (Off Duty): You need to protect yourself from the elements.	Example of an off-duty hat.  Courtesy of Brian Henington
Carabiner: Allows you to hook gloves, keys to your belt.	Used to hang clothes on belt loops and for securing other equipment.  Courtesy of Brian Henington

(continued)

<table>
<tr><td>Spanner Wrench: Important and essential tool for firefighters working with a fire engine or portable pumps.</td><td>Standard spanner wrench.

Courtesy of Brian Henington</td></tr>
<tr><td>Pocketknife or Multi-tool: Many uses and applications. A multi-tool can be used for almost any task on the fireline. If your loved one wants to buy you a gift for becoming a firefighter, ask for a multi-tool.</td><td>
Courtesy of Brian Henington</td></tr>
<tr><td>Extra Bootlaces: Leather bootlaces will break. Ensure you have extra sets in your fire pack to fix any issue with broken or damaged laces.</td><td>Leather shoelaces (extra).

Courtesy of Brian Henington</td></tr>
<tr><td>Flashlight: Fire camp is usually dark. A small flashlight is essential.</td><td>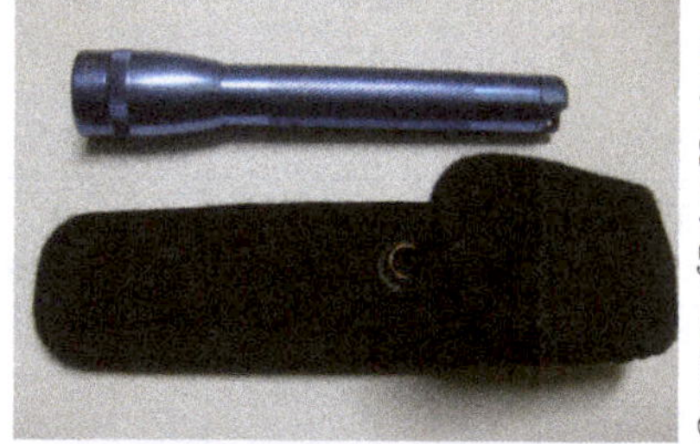
Courtesy of Brian Henington</td></tr>
<tr><td>Blue Pens: All original signatures must be in blue ink. Have numerous blue pens in your fire kit and overnight gear bag.</td><td>Blue pens—used for original signatures.
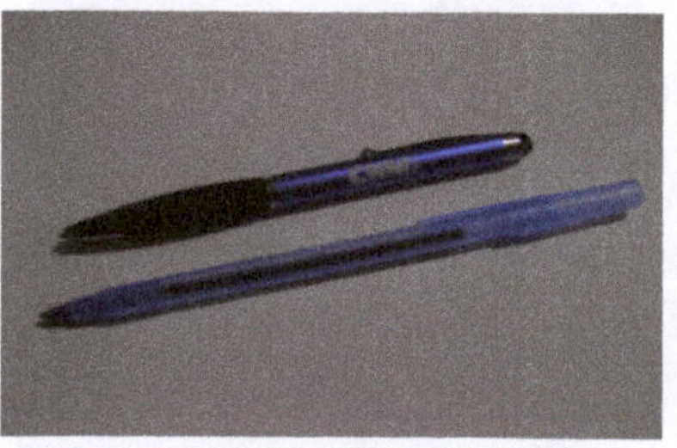
Courtesy of Brian Henington</td></tr>
</table>

(continued)

Tent: A tent allows you to change your clothes in private. Also keeps sleeping firefighters out of the elements.	Standard tent used by firefighters.  © Matee Nuserm/Shutterstock.com
Sleeping Bag: Ensure you have a sleeping bag rated for cold or cool weather.	Standard sleeping bag with compression outer bag.  © bogdan ionescu/Shutterstock.com
Jacket: You are in the mountains, so expect cool temperatures at night. You cannot wear a jacket on the fireline unless it is approved by NFPA.	 © Gingko/Shutterstock.com
Sunscreen Lotion: Sunburns are very common among firefighters. Carry in your fire pack and in your overnight bag.	 © DeiMosz/Shutterstock.com

(continued)

Lipbalm: Make sure your lipbalm has sun protection elements.	© Jeffrey B. Banke/Shutterstock.com
Matches or Lighter: Used to ignite firing devices. Matches or lighters should be kept in a fire protective case. There have been cases where lighters have exploded in the pockets of firefighters due to radiant heat. A wooden sheath is a good protection device for a lighter.	© Little_Desire/Shutterstock.com
Watch and Alarm: Purchase a cheap watch that has an alarm clock, back light, and military time (option). Alarm should be small and battery operated. Note: Do not count on your cell phone if you do not have adequate charging capabilities. You are responsible for getting up. It's not your supervisors' responsibility to ensure you wake up at the identified time.	© vilax/Shutterstock.com
Flip Flops: Athlete's foot is a major issue among firefighters. Firefighters share the same shower facilities. Carry a pair with you in your overnight bag.	© exopixel/Shutterstock.com
Towels: Do not rely on shower units to provide towels. If they do, they are usually small.	© sagir/Shutterstock.com

(continued)

Hand Sanitizer: Perfect when you are away from a hand-washing unit.	© kezza/Shutterstock.com
Pillow: Helps you get a good night's rest. Ensure it is small enough to fit in your overnight bag.	© Zonda/Shutterstock.com
Prescription Medicine: Enough for a 14-day assignment. Make sure your supervisor and crew EMT are aware of the medicine. If you are taking pain pills, you do not need to be fighting a fire.	© Glenn Price/Shutterstock.com
Extra Clothing: At least one extra fire shirt and fire pants. See Appendix A for a list of recommended items to have in your overnight bag.	
Personal Hygienic Gear: Ensure you have all the items needed for extended stay away from your house or fire department. See Appendix A for recommendations.	
Toilet Paper or Baby Wipes: Self-explanatory. Don't have it and find out how miserable life will become.	
Cash: Minimum of $50.00. Many stores in rural America do not take a credit or debit card.	

Weight of Gear

Entry-level firefighters are restricted to the amount of gear/equipment they can carry on any given assignment. This is especially true for firefighters working on a 20-person hand crew. The maximum total gear/equipment weight is 65 pounds. This includes 45 pounds of gear in your overnight bag (including sleeping bag) and 20 pounds in your fire pack (does not include water). Depending on what crew you work for; the standards will be strictly enforced. Ask a seasoned firefighter on how they pack their overnight bags and what gear is essential. You will discover techniques to pack your gear efficiently.

WEIGHT LIMITATIONS FOR YOUR GEAR

- Total weight—65 lbs.
- Overnight (personal gear)—45 lbs.
- Fire pack/web gear—20 lbs. (excluding water)

Accountability of Agency Equipment/PPE

Fire equipment and PPE are very expensive. Agencies equip firefighters with hundreds or even thousands of dollars' worth of gear/PPE. Fire pants range from $140.00 to $280.00. Hard hats are a minimum of $35.00. If you are assigned a radio, this could increase the cost by $2,000.00 to $3,000.00. It is your responsibility to maintain and secure the equipment provided to you. The equipment or PPE belongs to the agency. We must prioritize maintaining the equipment and PPE at the highest standard to maintain the most effective use.

Agencies are not responsible for personal items. If you provide your own gear, then the gear is yours; thus any damage or repair cost is on you, not the agency you work for. It is highly recommend that you use your agency or your fire department equipment and not your personal gear to avoid issues.

The environment we work in is dirty. We need to keep our equipment issued as clean as possible. The equipment provides a purpose. Introducing additional agents or chemicals on your gear will make them ineffective. Furthermore, after each assignment, make sure you clean your gear. Blow dust and ash off with an air compressor. Wash your fire-resistant clothes (make sure you follow washing instructions). Some firefighters pride themselves on not washing their clothes throughout the fire season. This attitude is not very smart. "Nomex™ loses its fire retardant capabilities if foreign substances are on or in the fibers."[3] Ripped clothing is also ineffective and should be replaced.

Teaching Note: If NFPA and OSHA requires specific PPE for wildland firefighters, than there is a reason. I have heard of firefighters sewing patches on their fire shirts. They were involved in a burnover and the only thing that burned was the patch. Do not add items to your gear that does not meet NFPA standards. If you want to look cool, then do it when you are off the fireline, not while you are fighting fires.

[3] National Wildfire Coordinating Group, *Firefighter Training: S-130* (Boise: National Wildfire Coordinating Group, 2003), 1.11.

FIREFIGHTER RESPONSIBILITIES

Firefighters are the key to extinguishing fires. We all have an important role to ensure the objectives of the incident are met. Regardless of what type of crew or resource you work with, we all have common job responsibilities that must be performed on every fire, every day, and throughout your career. The National Wildfire Coordination Center provides a clear list of job responsibilities. These are as follows:

1. Performs manual and semi-skilled labor.
2. Ensures the objectives and instructions are understood.
3. Performs work in a safe manner.
4. Maintains self in the physical condition required to perform the duties of fire suppression.
5. Keeps personal clothing and equipment in serviceable condition.
6. Reports close calls, accidents, or injuries to supervisor.
7. Reports hazardous conditions to supervisor.[4]

Courtesy of Brian Henington

Wildland Firefighter Miranda Miller.

Crew Organization

Crew organization will vary depending on what crew or agency/fire department you work for. Most crews value the importance of organization. Organization on a crew helps to ensure each firefighter understands his or her role and responsibilities. Crew organization establishes a defined and well-structured chain of command. Each firefighter should know exactly who their supervisor is and how the chain-of-command functions. Furthermore, a well-organized organization helps develop teamwork, professionalism, maturity, and respect for other crew members. Finally, accountability is created and enhanced and prioritized among organized crews. Your crew should function as a well-oiled machine and you should ensure that you play a role in maintaining its effectiveness.

Courtesy of Brian Henington

Crew members have to maintain organization at all times to ensure communication and safety measures are in place.

Ethics

Ethical behavior is a concern for all industries and occupations, not only wildland firefighting. A group of business professors and researchers has provided an excellent definition of ethics. According to Thomas Wheelen and J. David Hunger in their book *Strategic Management and Business Policy*, "Ethics is defined as the consensually accepted standards of behavior for an occupation, trade, or profession."[5]

[4] Ibid., 3.2

[5] Thomas Wheelen and L. David Hunger, *Strategic Management and Business Policy* (New Jersey: Prentice Hall, 2002), 43.

Furthermore, ethics can be described as the way we behave and act in relation to our defined and expected job duties. All eyes are on firefighters. Any unethical behavior or attitude will directly impact the public's perception of us.

Ethical issues do exist in wildland firefighting. A common activity is exaggerating the hours worked in an operational period. A firefighter works a 12-hour shift but on his crew time report he lists he worked 13.5 hours. This activity happens all the time. Be honest because somebody is eventually going to find out about your unethical behavior and the results or actions of your behavior are not going to be pleasant.

A good approach to understand if your actions are ethical is: If you have to justify the action to yourself, then it is more than likely unethical. According to the *International Fire Service Training Association*, in their textbook, *Fire and Emergency Services Instructor (8th Edition)*, the following justification measures are good indicators of unethical actions:

1. Pretending that the action is legal or ethical.
2. Believing that the action is really in the best interest of the organization or individual.
3. Believing that the action is OK because no one will ever discover it.
4. Expecting that the organization will support the action if it is ever discovered.
5. Believing that the action is acceptable because everyone else is doing it.
6. Believing that the end (result) justifies the means (method) even when the means are unethical.[6]

Professionalism

Courtesy of Brian Henington

Wildland Firefighters Erik Nelson and Diego Villalba.

Professionalism describes the behavior and attitude we possess as firefighters. It may include the agency's culture and mission, or it can include our own values and beliefs. Furthermore, a professional firefighter is one who acts in a manner that others see as honorable. The way we carry ourselves or the way we talk to others is an important part of professionalism. Professional firefighters should be trustworthy, honest, respectful, and accountable for their own actions.

Overall, firefighters are held in high regard by the public. Little kids aspire to become firefighters and look at us as heroes. Even though we do not consider ourselves as such, we need to remember that the public does. The number one way to change the public's perception about us is the way we act and talk in public. I was returning home from a long fire assignment (off-duty) and stopped in a convenience store. There were several firefighters in the store buying drinks and snacks. A little boy in the store was staring at the firefighters. He was whispering to his mom. I could only imagine what was being said between him and his mom. It made me proud of my profession. The little boy approached the firefighters, I assume to ask a question or give them a high five. As he approached, one of the firefighters dropped an "F bomb." It was loud enough for me to hear as well as the little boy. The boy stopped dead in his tracks.

[6] International Fire Service Training Association, *Fire and Emergency Services Instructor: 8th Edition* (Stillwater, Oklahoma: Oklahoma State University Fire Protection Publications, 2012), 25.

He turned around and ran back to his mom. His mom approached the firefighters and gave them a butt chewing. The firefighters never realized they did something wrong. Regardless if you believe that foul language is appropriate or not, this mom and boy did not. Did their image of us change?

Firefighters are tough. Wildland firefighters are even tougher! You have to be tough to do this job. You have to be willing and able to go without a shower for days on end. You sleep on the ground. You smell like body odor mixed with smoke. It's a dirty job! With that said, there is no excuse not to act as a professional. Your actions are a reflection on your agency. Fire managers expect people to act in a professional manner. Do not act in a manner or use words that would embarrass you are your family if they were ever made public. In today's society, people are capable of recording your actions and are quick to display any negative behavior on social media. Watch what you say and where you say it.

Appearance

Your appearance should reflect the standards established by your agency, fire department, or organization. Some agencies may have a specific dress code that includes a uniform for daily or non-suppression activities. The uniform represents the agency in a professional manner. The uniform is a symbol; it's a reflection of an honorable profession. Ensure you are wearing your uniform at the highest standard available.

On the other hand, wildland firefighters usually do not have an assigned uniform (except when they are on an active fire). What you should consider is how your non-fire clothes reflect on your agency, fire department, or organization. Do not report to work (even if it is mandatory physical training) in shirts or clothes that are not appropriate. Wear clothes that reflect positively on yourself. Shirts that say "Firefighters—We Find Them Hot and Leave Them Wet" are not appropriate apparel, as they are offensive to some and bring a negative reflection on your crew, your agency, and more importantly, yourself. Stay away from clothes that sponsor alcohol, tobacco, nightclubs, etc. Furthermore, do not wear anything that could be considered revealing or sexual that could be construed as harassment.

Most wildland fire crews will have their own crew tee-shirt. This builds pride and *esprit de corps* among the crew as well as provide a visual identification to the public. Wear these shirts. Look like a professional. An agency will never tell you what you can wear off duty, but when you are associated with fire activities, you need to represent your agency in the best manner possible.

Personal Hygiene

As mentioned throughout this chapter, we work in a very dirty environment. We work close to other firefighters as well as share close sleeping areas. The opportunity for germs or a cold to make the way from firefighter to firefighter is greatly increased. In addition, we are physically exhausted and our immune system is weak. The most common sickness is what firefighters call "Camp Crud." Camp crud spreads like wildfire among firefighters. Firefighters need to ensure we are making our best effort to reduce the exposure to germs. Frequently wash your hands or use hand sanitizer when washing stations are not available. Taking a shower when one is available is a must. Do not be the firefighter who thinks not taking a shower is a sign of toughness. It's not. Not cleaning yourself can lead to potential medical issues that will impact other firefighters. Use foot powder for your feet (to reduce athlete's foot) and in your groin (to reduce the potential for jock itch). It is also recommended that you supplement your diet with vitamins that help improve your immune system. Examples include Vitamin C, zinc, or multi-vitamins.

Courtesy of Brian Henington

Our fellow firefighter—Rest in Peace.

Cultural or Social Differences

It takes thousands of firefighters to fight fires across the United States. The National Fire Protection Association estimates there are approximately 1,140,750 firefighters in the United States.[7] We all come from different backgrounds, genders, races, ethnicities, religious beliefs, political affiliation, etc. I always tell my firefighters you do not have to believe in what I do, but you better respect the fact that I have the right to have my own beliefs, cultural values, and philosophies. I will not try to push those on others nor do I expect others to push their beliefs on me. The point is, we need to be respectful to others and their beliefs. A good way to get fired is make a comment that upsets or impinges on somebody else's values.

There is zero tolerance for any form of discrimination or harassment on the fireline or at the firehouse. Comments about race, age, sexual preference, religion, or values, as well as comments that could be taken as sexual harassment, only create dissention among the crews and can lead to complaints and/or termination. Be a professional. Treat others the way you want to be treated! Firefighters should further refrain from using racial terms at any time (this includes making references or using racial offensive words to members of your own race). If you constantly call one person of your race a racial charged term, people of other races will think that it is OK for them to do the same. Stay away from it. Don't create more issues than need be. Furthermore, stay away from any sexual related comments or jokes. Most fire managers give one chance. If you do not correct your actions, you need to find employment elsewhere. We will not tolerate insensitive, sexual, or racial comments at any point or to any extent. Remember, we are fighting fires. We need to focus and concentrate on the dangers and associated hazards. We do not need more distractions.

---

[7] Hylton J.G. Haynes and Gary P. Stein, National Fire Protection Association, *US Fire Department Profile 2013* (Maryland: National Fire Protection Association, 2014), iii.

Agency Affiliation

Courtesy of Brian Henington

Firefighters attending a briefing session prior to work activities.

Agency affiliation is best described as the agency that employs you and supports your firefighting efforts. You may work for a federal, state, county, tribal, local fire agency, volunteer fire department, or a private organization. The point is, fires demand large amounts of organizations and resources to suppress them. Have pride in your agency but understand that we are all here performing a duty to help put the fire out.

I have worked with every agency identified above. These agencies desire to hire and produce professional and competent firefighters. There is no reason for one agency to treat another agency with disrespect. This happens more than fire managers would like to claim. If somebody is not capable of what you are doing, teach them; don't disrespect them. Too many firefighters have treated others in a disrespectful manner because of their agency affiliation. My philosophy is this: We are here, risking our lives. I am not treating anybody badly because there equipment is not as good as ours. Help them to become better. Don't be disrespectful to others because of your agency affiliation. That attitude is a distraction to firefighter safety and presents discontent and anger among other crews.

Conflict Resolution

Conflict is highlighted in the *Fire Officer Principles and Practices (3rd Edition)* as, "a state of opposition between two parties."[8] Conflict or disputes are a common occurrence in any occupation. They may be heightened on the fireline because of the elevated stress levels and the life or death decision-making responsibilities we have. The successful firefighter has recognized that conflict will occur and understands the techniques on how to successfully resolve the issues. Dealing with conflict requires that all involved must first recognize there is an issue. You then need to define the problem and how can it be resolved. All parties should be listened to and the solution should be based on agency policy and the fairest and most reasonable solution.

CONFLICT RESOLUTION STEPS

1. Recognize or identify the conflict or dispute.
2. Address the dispute on a one-to-one basis (if you feel safe).
3. Remove your emotion from the conflict (the hardest to do!).
4. Discuss the conflict and develop solutions.
5. Initiate the solution.
6. Evaluate the action.
7. Determine methods to reduce the conflict in the future.

If you identify a potential problem, address it; do not let it fester or get worse. Be professional about it, but make sure that you address the issue with the individual(s) who may have created the problem or the conflict. More than likely, the issue will be handled at this initial stage. We are all tired and hungry and that often causes us to become

[8] Ward, *Fire Officer*, 224.

short-mannered or agitated (grumpy). If you immediately address an issue with somebody in a positive and professional manner, than the issue will probably be handled at this level. If the issue continues or the result of you conversation was negative, then the complaint should be addressed with your immediate supervisor. Of course, there may be times when your immediate supervisor is the one creating the issue. If so, you should address the issue with his or her supervisor, ensuring you follow the chain of command.

There are many proven techniques and concepts to deal with conflict. The intent of this chapter is not to explore every option; it is to increase your consciousness that conflict is common in a high-risk occupation and your success or the success of your crew is based on your willingness to address the issue and initiate productive and positive outcomes. Take college-level courses on conflict resolution. Read a book about solution options for conflict. If you plan on being a supervisor, a majority of your time and energy will be dealing with issues. If you are the one causing the issues, stop; it is not professional. If you like conflict, then go fight in the UFC; there is no place for continuous issues with problem starters in an arena as dangerous as ours.

YOUR PROFESSIONAL DEVELOPMENT

Mentor and Coach

Courtesy of Brian Filip

Wildland Firefighters James Bostian and Manual Chavez working with an experienced coach/ mentor—Brian Henington.

A mentor is a person who helps you achieve your goals and objectives though positive reinforcement and encouragement. "Mentoring is a relationship in which managers at the mid-points of their careers aid individuals in the first stages of their careers. Technical, interpersonal, and political skills can be conveyed in such a relationship from the more experienced to the less experienced person."[9]

Coaching is an essential skill for most mentors. We should all strive to develop new firefighters so our profession has a succession plan in place. The National Fire Protection Association defines coaching as "a method of directing instructing, and training a person or group of people with the aim to achieve some goal or develop specific skills."[10] Take time and train others to ensure that our profession is maintained and preserved at the highest standards.

[9] Robert L. Mathis and John H. Jackson. *Human Resource Management: Essential Perspectives* (Cincinnati: South-Western College Publishing, 1999), 94.

[10] Ward, *Fire Officer*, 142.

Teamwork

Courtesy of Brian Henington

Members of a crew working as a team. Teamwork is an essential and critical component in firefighting.

Teamwork is the key component to successful and efficient fire crews. Regardless of the type of resource you work on, teamwork is essential and must be developed and maintained. The bottom line in our dynamic, challenging, and complex environment is you have to count on your brothers and sisters to keep you out of dangerous situations. You also have to count on them to complete their individual job tasks so the overall mission of the crew is accomplished. Good supervisors (who prioritize the team over themselves) will continually improve the overall team concept within their crew. This philosophy ensures the most efficient and effective resources are available for assignment. The great National Football League coach, Vince Lombardi, summarized teamwork as "People who work together will win, whether it be against complex football defenses, or the problems of modern society." Phil Jackson, the legendary championship NBA coach, also summarized teamwork as: "The strength of the team is each individual member…the strength of each member is the team."

Good fire managers focus on the concept of teamwork. Without it, we are just individuals pursuing our own self-interest. This attitude will create dissention and disrespect among a crew. Crews and supervisors should strive to develop and maintain the best crew through prioritizing the concept of team as it relates to safe and efficient firefighting operations. Teamwork fosters positive and effective communications. You cannot be a safe firefighter if you do not have adequate communication among the crew members and supervisors. You should prioritize the crew and team above yourself. Remember, in the eyes of most fire managers, a good team player will become a good leader based on his or her ability to make others on their team the top priority.

The National Wildfire Coordinating Group provides several recommendations for developing and maintaining successful teamwork among firefighters. They include:

1. Know yourself and seek improvement.
2. Be technically and tactically proficient.
3. Comply with orders and initiate appropriate actions in the absence of orders.
4. Develop a sense of responsibility and take responsibility for your actions.
5. Make sound and timely decisions and recommendations.
6. Set the example for others.
7. Be familiar with your leaders and their jobs, and anticipate their requirements.
8. Keep your leaders informed.
9. Understand the task and ethically accomplish it.
10. Be a team member but not a "Yes Person."[11]

[11] National Wildfire Coordinating Group, *Firefighter Training: S-130* (Boise: National Wildfire Coordinating Group, 2003), 4D.9.

TABLE 6-1 *THE DIFFERENCE BETWEEN EFFECTIVE TEAMS AND BAD TEAMS*

Successful Teams	Bad Teamwork
• Good attitude is maintained through all ranks of the team. • Communication is prioritized and maintained at all times. • Supervisor prioritizes the crew's interest above his or her own aspirations. • Crew members prioritize a high work ethic and technical proficiency. • Crews can function without direct and immediate supervision. • Everybody focuses on improving his or her roles and abilities. • The team (crew) is number one. Pride is built and maintained.	• Bad attitude is obvious among individuals or crew members. • Laziness. Crew members not pulling their weight and performing their specific job duties to the best of their ability. • Crews or crew members acting in an arrogant or with a "know it all" attitude. • Treating other crews or firefighters in a disrespectful manner. • Cliques among crew members. • Unethical activities. • Not taking responsibility for individual's mistakes or failures. • Passing the mistakes made to other crew members or crews.

FITNESS AND HEALTH

A firefighter's ability and willingness to maintain top physical conditioning is an essential element to this occupation. Being in shape reduces the strain on your body while also reducing the potential for injury. You should anticipate the requirements this job to be more than many other occupations, including athletic endeavors.

The Pack Test

The national physical fitness testing standard for wildland firefighters is the **Pack Test** or Work Capacity Test. The Pack Test must be taken annually, and all firefighters actively engaged in the suppression efforts of a fire must pass the test at the arduous level. The test requires firefighters to be in good physical conditioning. The pack test consists of a three-mile hike with a 45-pound pack or vest. The participant cannot run or jog; they must walk. They must complete the test (depending on elevation) within 45 minutes. Struggling through the test is an indication that you are not in the best shape to perform your job duties at the highest and most efficient level. Some agencies require medical clearances by a qualified medical professional before you can take the test. It is very unfortunate, but almost every year there is a fatality or serious injury associated with the firefighters taking the Pack Test.

PACK TEST BASICS

1. 3 mile hike (no jogging/running)
2. 45 lb. pack or vest
3. 45 minutes to complete (may vary depending on elevation)
4. Must be administered by a Pack Test Administrator
5. Medical Clearances are required for some agencies

Fireline Fitness

Firefighters should consider themselves athletes. Just as athletes are constantly improving their strength, quickness, and endurance, we too should be doing the same. Our job requires movements and physical activities that the general public does not perform in their jobs. We need to be in top physical shape. The better conditioning we are in, the easier our job will be. Wildland firefighting is extremely tough on the body. The wear and tear from the physical requirements of the job can lead to injuries, some of which may be extensive or severe.

Fatigue is a major concern in our profession. Fatigue can lead to injuries and it is considered a distraction to our situational awareness. To reduce fatigue, the best approach a person can have is to keep his or her body in ideal physical conditioning. The physical fitness program that my college supports for wildland firefighters is a program championed by a pioneer in wildland firefighting. The program, "FireFit," was produced by a task force of professional firefighters with the chairperson being a college professor and pioneer in wildland firefighting, Ms. Bequi Livingston. The program was developed to focus the attention of firefighters on the necessity of an elaborate, science-based approach to fitness. The program was specifically designed for the wildland firefighter. FireFit is explained in detail as, "FireFit is an easy-to-follow fitness program specifically designed for wildland firefighters and fire support personnel. Promoting year-round fitness and wellness, it provides the information and tools necessary for wildland fire personnel to develop their own personalized fitness program while minimizing the occurrence of injuries."[12] **Fire Fit** or **Fireline Fitness** strives to condition wildland firefighters to work out throughout the year. It is intended to change the way you view working out and how being in shape can and will help you perform your job duties at the highest and safest level possible. The program focuses on the concepts of aerobic fitness, strength and power, muscle endurance, core and stability and flexibility. The program also uses several fitness assessment evaluation criteria to gauge your development. If you use the program, you will find your body reshaping and becoming a top performing machine.

© Bequi Livingston

Fireline Fitness students working at Central New Mexico Community College.

Some firefighters have a very aggressive and fine-tuned work out regime. They are in top physical conditioning and are able to perform their jobs under very stressful and demanding conditions at the highest level. It is important that we are all aware that one of the top line-of-duty causes of fatalities for firefighters is medically related with a majority being heart attacks. (One hundred and thirteen fatalities related to heart attacks occurred of 2013.)[13] You need to prepare yourself for a physically demanding job and build your workout regime as such.

---

[12] Bequi Livingston, FireFit Task Group, National Wildfire Coordinating Group, *Are you FireFit? Interagency Wildland Firefighter Fitness Program* (Canada National Wildfire Coordinating Group: Quick Series, 2011) 1.

[13] National Interagency Fire Center, *Historic Wildland Fire Fatality Reports* (Boise: National Interagency Fire Center, 2014), 22.

Author's Perspective: A major issue with wildland firefighters and major issue with this author throughout his career was to work out as hard as you could two to three months prior to fire season. After fire season, our bodies are worn down. The common behavior was to rest your body until after the New Year. Then begin your workouts to get ready for the Pack Test. From experience, I would always get a minor injury that would slow down my ability to work out. I would pass the Pack Test (and hate every second of it), then be ready for fire season. The first fire at high elevation would test my body's conditioning. The older I get, the more I realized that this type of workout regime is the wrong regime. I have started using the Fireline Fitness program instructed at my college. Fireline Fitness is equivalent to the workouts I use to have as a college football player. I have refocused my attention to the necessity of physical conditioning and have personally witnessed a change in my body, my conditioning, strength, and endurance. I am healthier now because of this program.

Diet

Diet is also as important as your physical conditioning. There are numerous experts and publications on the subject of diet and nutrition. It is recommended that you follow a solid diet plan that includes avoiding certain food or drinks. Again, we highly recommend following the diet component under the FireFit or Fireline Fitness program.

Some food or drinks that you should be concerned with during fire season:

1. Alcohol: Alcohol dehydrates the body and becomes an impairment to judgment. You should avoid alcohol as much as possible during fire season. Alcohol also impacts your body's ability to reach top performance during workout sessions.
2. Energy drinks: More studies are focusing on the negative side effects of energy drinks. Energy drinks create a quick burst of energy, but a major side effect is crashing. Crashing is a distraction to situational awareness; thus energy drinks are not recommended during fire suppression activities.
3. Caffeine: Coffee and tea, as well as most soft drinks act as a diuretic that can lead to dehydration, which may lead to a heat-related illness. Limit your caffeine intake during fire activities.

Fatigue

Fatigue is a problem among firefighters. Working long hours without adequate rest will greatly increase your fatigue levels. Being in top physical conditioning will help reduce fatigue; however, adequate rest is the best medicine. Fatigue develops because we normally work 12 to 16 hour shifts, up to 14 days consecutively. Fatigue is another distraction to situational awareness. Your brain will not work properly when the body is deprived of rest. Seasoned firefighters have learned techniques to help deal with fatigue. Learn from them on what you can to reduce the impact fatigue has on your body. You should also ensure you are following the standard work to rest ratio: two hours worked for one hour rest. Sixteen hours

NATIONAL WILDFIRE FIREFIGHTING STANDARD

Work to Rest Ratio:

2 hours worked = 1 hour of rest

16 hours worked in a day = 8 hours of rest

worked equate to eight hours of rest. Make sure you are using your rest periods in the most appropriate manner. Get the rest your body deserves.

Hydration

The most important thing you can do for your body is to drink water. Staying hydrating is significantly important to firefighters. Extreme temperatures, sweating, and improper diet will cause the body to lose massive amounts of water. Monitor your water intake. A helpful tool to help you monitor dehydration is your urine. If you are not urinating normally or on the hour, it is a good indication that dehydration is setting in. In addition, if your urine is dark in color, you may also be experiencing early signs of dehydration. Signs of dehydration include headaches, lethargic feeling, constipation, and/or irritability.

Dehydration can lead to a heat related injury or illness. These injuries or illnesses include: 1) heat cramps, 2) heat exhaustion, and 3) heat stroke. The *Incident Response Pocket Guide* provides a very useful reference for symptoms and treatment options for the three injuries described above. The table below provides a summary:

TABLE 6-2 ***HEAT-RELATED INJURIES***

Type of Injury/Illness	Symptoms/Signs	Treatment Options
Heat Cramps	• Sweating • Dehydration • Transient muscle cramps	• Place in shade • Loosen clothing and stretch muscles • Slowly give fluids • Monitor
Heat Exhaustion	• Profuse sweating with cool, clammy skin • Dehydration • Persistent muscle cramps • Dizziness and headache • Decreased urine output	• Place in shade • Loosen clothing and stretch muscles • Slowly give fluids • Monitor; medevac if no improvement
Heat Stroke	• Hot, dry skin • Rapid, weak pulse (100–120 at rest) • Hyperventilation • Vomiting • Involuntary bowel movement • Dizziness, confusion, and irritability • Seizures or loss of consciousness	• Cool body as quickly as possible with water (river, fold-a-tank, canteens, etc.) • **MEDEVAC IMMEDIATELY**

Information provided by the National Wildfire Coordinating Group. *Incident Response Pocket Guide 2014.*

SUMMARY

The requirements to participate in wildland fires are extensive. The requirements have been put in place to reduce the potential for injury while increasing a safe work environment. Personal protective equipment must be worn during all fire suppression activities to protect you and other firefighters. We must be accountable for our equipment and ensure it is functioning at the most effective level. As firefighters, we must also be ready for a dispatch on a second's notice. Having your gear ready is an important step for all entry-level firefighters.

Firefighters must also understand core essentials needed to function as a professional firefighter. Understand the importance of professionalism, teamwork, cultural diversity, and conflict resolution. If you use the concepts identified in this chapter, you will have a positive career as a wildland firefighter. Finally, if you understand how important fitness and diet are to firefighters, you will decrease the risk of serious injury and improve your ability to function at a high and efficient level.

KNOWLEDGE ASSESSMENT

1. List the standard personal protective equipment required to participate in wildland fire activities.
2. Explain the importance of accountability and maintenance of personal protective equipment.
3. What are the minimum standards for a wildland firefighter's boot?
4. What are the national standard weight limitations for a firefighter's gear?
5. Name four of the seven firefighter responsibilities.
6. Why is it important to maintain professionalism?
7. Explain cultural differences and how they impact wildland fire activities.
8. What are three steps involved with successful teams?
9. What are three activities common with unsuccessful teams?
10. What does the Pack Test consist of?
11. Is fitness important to the wildland firefighter?
12. What are the three heat related injuries/illnesses that can affect firefighters?

EXERCISES

1. Research, develop, and plan for a rigorous workout or fitness program. Start immediately so you can ensure you are in the best possible shape to handle the rigors of this job.
2. Using your *Incident Reponses Pocket Guide*, locate and familiarize yourself with heat-related injuries. Have you experienced any of the systems or injuries before? Select one type of heat-related injury and list for your instructor the symptoms and treatments and research one incident that involved a heat-related injury. (Provide reference source and type of injury; it does not have to involve a firefighter).
3. Before you get your first fire assignment, we want you to consider these situations that impact wildland firefighters. The purpose of this exercise is to have a plan in place so you can implement it once you get a fire assignment. 1) Have you made arrangements for somebody to pay your bills, water your lawn, etc.? 2) Do you have arrangements for your pets? 3) Have you told your loved ones that you love them?

BIBLIOGRAPHY

Haynes, Hylton J.G., and Gary P. Stein. *US Fire Department Profile 2013*. Maryland: National Fire Protection Association, 2014.

International Fire Service Training Association. *Fire and Emergency Services Instructor: 8th Edition*. Stillwater, Oklahoma: Fire Protection Publications, Oklahoma State University, 2012.

Livingston, Bequi. FireFit Task Group. *Are you FireFit? Interagency Wildland Firefighter Fitness Program*. Canada National Wildfire Coordinating Group: Quick Series, 2011.

Mathis, Robert L., and John H. Jackson. *Human Resource Management: Essential Perspectives*. Cincinnati: South-Western College Publishing, 1999.

National Fire Protection Association, NFPA 1977. *Standard on Protective Clothing and Equipment for Wildland Firefighting: 2011 Edition*. National Fire Protection Association, 2011.

National Interagency Fire Center. *Historic Wildland Fire Fatality Reports*. Boise: National Interagency Fire Center, 2014.

National Wildfire Coordinating Group. *Firefighter Training: S-130*. Boise: National Wildfire Coordinating Group, 2003.

National Wildfire Coordinating Group. *Incident Response Pocket Guide—2104 Version*. Boise: National Wildfire Coordinating Group, 2014.

United States Department of Labor: Occupational Safety & Health Administration. *Definition of Personal Protective Equipment*. United States Department of Labor: Occupational Safety & Health Administration. https://www.osha.gov/SLTC/personalprotectiveequipment/.

Ward, Michael J. *Fire Officer: Principles and Practices: 3rd Edition*. National Fire Protection Association: NFPA. International Association of Fire Chiefs. Burlington, MA: Jones & Bartlett Learning, 2015.

Wheelen, Thomas, and L. David Hunger. *Strategic Management and Business Policy*. New Jersey: Prentice Hall, 2002.

CHAPTER 7

WILDLAND FIRE RESOURCES AND TRANSPORTATION SAFETY MEASURES

Learning Outcomes

- Identify the difference between a single resource, strike team, and task force.
- List and identify the common resources used on a wildland fire.
- Explain how resources are categorized and defined by established national standards.
- Express the national chainsaw certification standards used on wildland fires.
- Identify the driving limitations that firefighters must adhere to while on assignment.
- Recognize that driving a fire vehicle can lead to serious accidents or fatalities.
- List the safety procedures firefighters should follow while flying on a helicopter.

Courtesy of Brain Henington

Key Terms

Air Tankers
Dozers
Driving Limitations
Engines
Firing Crews
Hand Crews
Heavy Equipment
Helitack
Interagency Hotshot Crews
Resource
Resource Kind
Resource Type
Retardant
Sawyers or Fallers
Single Resource
Smoke Jumpers
Strike Team
Swamper
Task Force
Transportation Safety
Water Tenders

OVERVIEW

A wildland fire resource involves many types of equipment, crews, and/or individuals. The resources work together to perform a specific function or job in order to suppress a fire. The resources must function in a safe and efficient manner in order to ensure the best possible outcome is achieved. Resources are classified based on their kind and type. The kind of resource identifies what type of equipment it is, while the type summarizes the capabilities of the resource. Resources can be grouped together to form a team to ensure more effective operations or activities. This chapter will introduce you to several resources used on wildland fires and provide a summary of each.

Driving to and from a fire can present numerous safety concerns or challenges. This chapter will further introduce you to driving safety concerns and provide a summary of how many accidents have occurred while operating a fire vehicle. Furthermore, when you arrive on a fire, you may have to use additional transportation methods to access your work location or assigned area. You will become familiar with the safety standards associated while walking to your assignment, riding in a helicopter, or transported by a boat.

WILDLAND FIRE RESOURCES

A wildland fire resource can include any of the following: a firefighter, fire engine, hand crew, aircraft, or a piece of heavy equipment. It can also include specialized assets that perform specific functions like helitack, firing crews, felling crews, or air tankers. The National Wildfire Coordinating Group clearly defines a **resource** as, "Personnel, equipment, services and supplies available, or potentially available, for assignment to incidents. Personnel and equipment are described by kind and type, e.g., ground, water, air, etc., and may be used in tactical, support or overhead capacities at an incident."[1] Resources can work independently or assigned as a specific group created to manage a specific task.

Wildland fire resources are classified into two categories based on their capabilities and function or purpose. The first class is a **resource kind**. According to the Federal Emergency Management Agency, a resource kind "describes what the resource is…"[2] On wildland fires, resources kinds are classified into fire engines, hand crews, aircraft, heavy equipment, bulldozers (called dozers), helitack, firing crews, falling crews, etc. Fire managers further classify each resource kind into specific categories.

A **resource type** specifies the capabilities or function of the specific kind (example—fire engine). A resource type "describes the size, capability, and staffing qualifications of a specific kind of resource."[3] The *Fireline Handbook* further describes resource types as "Resource typing provides managers with additional information in selecting the best resource for the task."[4] The ranking of Type 1 associated with a specific kind of resource usually identifies the resource as the most capable, prepared, and equipped resource available for assignment. However, this is not always the case. For example, a Type 1 engine is a structural fire engine. A Type 6 engine is specifically designed for wildland fire operations and has better capabilities off-road than a Type 1 engine would, thus making it more suitable for wildland fire operations.

---

[1] National Wildfire Coordinating Group, *Glossary of Wildland Fire Terminology* (Boise: National Wildfire Coordinating Group, 2014) 149.

[2] Federal Emergency Management Agency, *IS-703.A: NIMS Resource Management* (United States Department of Homeland Security, 2010) 4.4.

[3] Ibid., 4.4.

[4] National Wildfire Coordinating Group, *Fireline Handbook* (Boise: National Wildfire Coordinating Group, 2004), 350.

The type and kind of resources needed to suppress a fire depends on the complexity, logistics, and special circumstance specific to each incident. Most resources are assigned to a fire as a single resource with a specific function or duty. However, when a fire transitions to large scale or longer duration event, then fire mangers will group single resources into strike teams or task forces. This philosophy allows for better management of resources through close supervision, direction, improved communication, and more effective utilization of the resource.

A **single resource** is described as "An individual, a piece of equipment and its personnel complement, or a crew or team of individuals with an identified work supervisor that can be used on an incident."[5] The engine you are assigned to (Engine 42) is considered a single resource. If you work for the Santa Fe Hotshots, then you are still considered a single resource.

Courtesy of Brian Henington

Strike Team of Type 6 Engines.

When resources are assigned to work together, then the result is either a strike team or a task force. The National Wildfire Coordinating Group describes a **strike team** as "specified combinations of the same kind and type of resources, with common communications, and a leader."[6] Strike teams must have the same type and kind of resource. Strike teams can vary from two to seven resources (similar type and kind) with the ideal component involving five resources with one assigned supervisor. The most common strike teams involve fire engines; however, it is not uncommon in certain parts of the United States to see strike teams of dozers or hand crews. An example of a Strike Team would be five Type 6 Engines. The team would have one supervisor (Strike Team Leader) and would be assembled to improve use and effectiveness of the five engines that work together to accomplish an objective.

Once resources are mixed, based on kind and/or type, the situation would be called a **task force**. A task force is defined as "any combination of single resources assembled for a particular need, with common communications and a leader. A task force may be pre-established and sent to an incident, or formed at an incident"[7] The task force involves either the same kind or

EXAMPLE OF AN INITIAL ATTACK TASK FORCE

(2) Type 6 Fire Engines: (quick mobile attack, able to get into almost any location).

(2) Type 3 Fire Engines: (more water and pumping capabilities, more firefighters, and better foam capabilities).

(1) Initial Attack Hand Crew—10 firefighters: (capable of structure triage and protection, chainsaw, and firing operations).

(1) Tactical Water Tender: (provides water to engines, but can be used to support grass fires as an additional engine).

(1) Support Water Tender: (provides immediate water to firefighting operations—many parts of the country do not have available water for suppression activities).

[5] National Wildfire Coordinating Group, *Glossary of Wildland Fire Terminology*, 158.

[6] Ibid., 157.

[7] National Wildfire Coordinating Group. *Fireline Handbook* (Boise: National Wildfire Coordinating Group, 2004) 350.

different kinds of resources, but they will be of different or multiple types. An example of a task force is two Type 6 Fire Engines, two Type 4 fire engines, one dozer, and one water tender. Another example would be three Type 6 fire engines and two Type 3 fire engines. Even though the fire engines are the same kind of resource, they are different types; thus making them a task force and not a strike team. The task force is supervised by a task force leader. The concept of task forces is becoming more popular because of the increased capabilities and function of the individual resources associated with the team.

Fire Engines (from this point forward referred to as engines)

Engines will consist of firefighters using a fire engine to provide and pump water, develop hose lays, and/or construct fireline. Engine crews are well versed and may also assist with structure protection activities, firing operations, chainsaw tasks or other specialized jobs. The individual in charge or responsible for the engine and crew is called an Engine Captain or Engine Boss. Note: The capabilities or minimum specifications listed below are national standards that must be achieved in order for the engine to participate on a wildland fire.

TABLE 7-1 *TYPE 1 ENGINE*

Specifications: Structure fire engine. Best use is structural protection. Must have available water source (fire hydrants).

# of Firefighters	Tank Size	Pumping Capability	Hose 2 ½″	Hose 1 ½″	Hose 1″
4	300 gallons	1,000 GPM	1,200 ft.	500 ft.	N/A
Pump and Roll Capabilities	No		**Best Use**	Structure Protection	

© Brad Sauter/Shutterstock.com

- GPM—Gallons Per Minute
- Pump and roll identifies an engine's capability to move with the fire or move quickly. This technique relies on the engine's own water supply, and not water from a fire hydrant. This technique is highly effective in wildland fire especially in light fuels.

TABLE 7-2 *TYPE 2 ENGINE*

Specifications: Structure fire engine. Best use is structural protection. Must have available water source (fire hydrants). Has less pumping capability than a Type 1 Engine.

# of Firefighters	Tank Size	Pumping Capability	Hose 2 ½″	Hose 1 ½″	Hose 1″
3	300 gallons	500 GPM	1,000 ft.	500 ft.	N/A
Pump and Roll Capabilities	No		**Best Use**	Structure Protection	

© mikeledray/Shutterstock.com

- GPM—Gallons Per Minute

TABLE 7-3 *TYPE 3 ENGINE*

Specifications: More common wildland fire engine. Best pumping capability of all wildland engines. Usually has powerful central air foam system. Preferred for pumping water to long or difficult hose lays.

# of Firefighters	Tank Size	Pumping Capability	Hose 2 ½″	Hose 1 ½″	Hose 1″
3	500 gallons	150 GPM	N/A	1,000 ft.	500 ft.
Pump and Roll Capabilities	Yes		**Best Use**	Wildland Fire Activities	

© Jim Parkin/Shutterstock.com

- Pump and roll identifies an engine's capability to move with the fire or move quickly. This technique relies on the engine's own water supply, and not water from a fire hydrant. This technique is highly effective in wildland fire especially in light fuels.

TABLE 7-4 *TYPE 4 ENGINE*

Specifications: Has the largest tank capacity of all fire engines. Has become more popular in recent years.

# of Firefighters	Tank Size	Pumping Capability	Hose 2 ½″	Hose 1 ½″	Hose 1″
2	750 gallons	50 GPM	N/A	300 ft.	300 ft.
Pump and Roll Capabilities	Yes		**Best Use**	Wildland Fire Activities	

Courtesy of Brian Henington

- Pump and roll identifies an engine's capability to move with the fire or move quickly. This technique relies on the engine's own water supply, and not water from a fire hydrant. This technique is highly effective in wildland fire especially in light fuels.

TABLE 7-5 *TYPE 5 ENGINE*

Specifications: Good for mobile attack. Common engine used by the Montana Department of Natural Resources.

# of Firefighters	Tank Size	Pumping Capability	Hose 2 ½″	Hose 1 ½″	Hose 1″
2	400 gallons	50 GPM	N/A	300 ft.	300 ft.
Pump and Roll Capabilities	Yes		**Best Use**	Wildland Fire Activities	

Courtesy of Brian Henington

TABLE 7-6 *TYPE 6 ENGINE*

Specifications: Most common wildland fire engine. Often referred to as the work horse. The engine is very mobile and can get into almost any location. Usually built on a heavy-duty pickup chassis. Referred to as a brush truck by most structure fire departments.

# of Firefighters	Tank Size	Pumping Capability	Hose 2 ½″	Hose 1 ½″	Hose 1″
2	150 gallons	50 GPM	N/A	300 ft.	300 ft.
Pump and Roll Capabilities	Yes		**Best Use**	Wildland Fire Activities	

Courtesy of Brian Henington

TABLE 7-7 *TYPE 7 ENGINE*

Specifications: Very common for prevention patrols. This type of engine is also used by some Hotshot superintendents. Provides quick, but limited, water use. Usually supported by other types of engines.

# of Firefighters	Tank Size	Pumping Capability	Hose 2 ½″	Hose 1 ½″	Hose 1″
2	50 gallons	10 GPM	N/A	300 ft.	N/A
Pump and Roll Capabilities	Yes		**Best Use**	Wildland Fire Activities—limited water supply	

Courtesy of Brian Henington

Hand Crews

Interagency Hotshot Crews—IHC: (Type 1).

Hand crews are an essential and vital resource for wildland fire activities. The resource provides a wide variety of use and skills. The *Fireline Handbook* defines hand crews as "a number of individuals that have been organized and trained and are supervised principally for operational assignments on an incident."[8] Hand crews can range from 4 to 22 firefighters, but the standard crew configuration is 18 to 22 firefighters. Hand crews specialize in fireline construction, chainsaw operations, and firing operations. Hand crews are also separated by type. The types include: Type 1 Interagency Hotshots, Type 1, Type 2 Initial Attack, and Type 2 Hand Crews.

By Agency

US Forest Service
Bureau of Land Management
Bureau of Indian Affairs
National Park Service
State Agency
County Fire Department/Agency

1
3
2
6
11
89

Information gathered from the US Forest Service Intergency Hotshop Crew webpage.

Figure 7-1 *Interagency Hotshot Crews by Agency.*

Interagency Hotshot Crews are deemed the most efficient, effective, and best trained hand crews. They are elite crews that deal with the most difficult tactical operations and assignments. Hotshot crews receive extensive training and are typically the most experienced hand crews on the fireline. In addition, these crews maintain rigorous physical fitness requirements due to the type of terrain and topography they work in. These crews are considered national resources whereas other crews may not have this consideration.

According to the United States Forest Service, there are 113 Interagency Hotshot Crews located throughout the United States.[9] They are assigned to a specific location and region, but they can be dispatched to any area experiencing high fire activity. California hosts the most Interagency Hotshot Crews (48) with the Southwest Region (Arizona/New Mexico) hosting the second most (19). The United States Forest Service employs the most hotshots; however, other federal, state, and county agencies also employ hotshots. See Figure 7-1 for agency affiliation for Interagency Hotshot Crews.

---

[8] Ibid., 334.

[9] United States Forest Service, *National Interagency Hotshot Crews (IHC)*, accessed February 14. 2015, http://www.fs.fed.us/fire/people/hotshots/IHC_index.html.

TABLE 7-8 INTERAGENCY HOTSHOT CREW MINIMUM SPECIFICATIONS

Crew Size	18–22	Leadership Qualifications (minimum)	• Task Force Leader • Incident Commander Type 4 • Firing Boss	Experience Levels	1 season or more must be 80% of crew
Full-Time Organized Crew	Yes	Minimum # of Full Time Employees	7	Communications	5 programmable radios (minimum)
Chainsaw Operations (Sawyers)	Minimum of 3 certified	Transportaion	Own transportation	Self-Supported	Yes—24 hours minimum
National Resources	Yes	Tools & Equipment	Self-Supplied	Special Skills	• Firing Operations • Aircraft Support • Chainsaw • Water and Pump Use

Type 1 Crews The classification of Type 1 Crews compared to Type 1 Interagency Hotshot Crew can be confusing for entry-level firefighters. An example of a Type 1 Hand crew would be 18 to 20 Smokejumpers that are assigned to an incident as a hand crew, not as resources that accesses the fire through the air. Type 1 crews will still maintain high experience and training standards; however, they may not train and work together on non-fire assignments (Interagency Hotshot Crews must train and work together a minimum of 40 hours a week when not on assignment).

Courtesy of Brian Henington

Type 2 Initial Attack Crews The recent trend among fire agencies is the creation and utilization of Type 2 Initial Attack Crews (Type 2IA). Type 2IA crews are not as qualified as Type 1 crews; however, they do possess chainsaw capabilities, which typically distinguish them from a standard Type 2 crew. Some Type 2IA crews provide their own transportation, chainsaws, and hand tools. The main difference between a Type 2IA crew and Type 2 crews is they can be separated into modules to support initial attack activities. They were originally created to support engine activities with most crews ranging from 4 to 10 members. These crews usually have experienced firefighters with advanced training and experience in burnout operations, chainsaw activities, and air resource coordination. The leadership of the crew must be qualified as a Crew Boss or higher. The crew usually has at least one Type 4 Incident Commander but must have a minimum of three Type 5 Incident Commanders.

TABLE 7-9 *TYPE 2 INITIAL ATTACK CREW MINIMUM SPECIFICATIONS*

Crew Size	18–20	**Leadership Qualifications (minimum)**	• Crew Boss • (3) Incident Commander Type 5	**Experience Levels**	1 season or more must be 60% of crew
Full-Time Organized Crew	No	**Minimum # of Full Time Employees**	0	**Communications**	5 programmable radios (minimum)
Chainsaw Operations (Sawyers)	Minimum of 3 certified	**Transportaion**	Varies depending on crew	**Self-Supported**	Varies depending on crew
National Resources	No	**Tools & Equipment**	Varies depending on crew	**Special Skills**	• Chainsaw • Firing Operations

Type 2 Hand Crews Type 2 Hand Crews are often referred to as seasonal or pick up crews. Type 2 Crews are usually assembled during on-going fire incidents and do not have as extensive training or experience that Type 1 or Type 2IA crews have. They may have their own transportation; however, most crews will need to be transported to the incident and equipped with tools. Most Type 2 crews do not have chainsaw capabilities. The management of the crew requires a Crew Boss and three Squad Bosses. Examples of Type 2 Crews include: Southwest Firefighters (SWFF); Montana Indian Firefighters (MIFF), Snake River Valley, agency regulars, Alaskan Native Crews, and private contractor crews.

TABLE 7-10 *TYPE 2 CREW MINIMUM SPECIFICATIONS*

Crew Size	18–20	**Leadership Qualifications (minimum)**	• Crew Boss • (3) Squad Bosses	**Experience Levels**	1 season or more must be 20% of crew
Full-Time Organized Crew	No	**Minimum # of Full Time Employees**	0	**Communications**	4 programmable radios (minimum)
Chainsaw Operations (Sawyers)	None	**Transportaion**	Needs transportation— in most cases	**Self-Supported**	No
National Resources	No	**Tools & Equipment**	Tools may have to be supplied	**Special Skills**	• Varies

Inmate Hand Crews Several western states use minimum-security inmates to support wildfire activities. These programs (usually referred to as Inmate Work Crews) provide a vocational skill for incarcerated individuals. The programs are designed to supplement fire crews by providing more resources to combat fires. Most programs also have a natural resource component (such as thinning trees or prescribed fire). Most inmate crews have skilled and experienced chainsaw operators due to their work on natural resource projects throughout the year. You should expect to work with or adjacent to inmate crews at some point in your career. A majority of the crews are very respectful, hardworking, and willing to support fire activities at their highest capabilities. The *Wildland Fire Incident Management Field Guide* provides specific rules when working with inmates.

Courtesy of Brian Henington

Type 2IA Inmate Crew: Crew boss is wearing a white hard hat. Inmates are typically identified by orange or red Nomex™.

Working with Inmate Fire Crews[10]

- Crews on fireline are supervised by forest crew supervisors (resource boss or higher).
- Inmate crews are usually limited to use within the state where they are based, although some states have interstate agreements with neighboring states.
- Contact with inmates should be done through the corrections officer-in-charge in camp.
- Contact with inmates should be done through the forest crew supervisor on the fireline.
- Consult the officer-in-charge before giving direction to inmates.
- Keep relationships with inmates on a business basis. For example, do not play cards with, carry messages for, bring gifts to, accept gifts from, or make purchases for the inmates.
- The officer-in-charge or other inmate camp representative may act as liaison with Fire Overhead on all matters pertaining to inmates (food, bedding areas, etc.).
- The officer-in-charge will remain with the crew while on the fireline. Any fire suppression related problems, such as pumps, tools, drinking water, fire equipment, etc., are to be taken care of by Fire Overhead.
- Inmates should not be used in a "Squad Boss" type position, or given supervision over fellow inmates.
- Inmate crews should be provided a separate sleeping area where they can be away from other crews.
- Provide separate sleeping areas for male and female, and adult and juvenile, crews.
- Interspersing inmate crews with civilian crews on the fireline is generally permitted (but not encouraged), provided the crew supervisor is aware of the situation at all times.
- Intermingling of inmates at the incident base with civilians should only occur at meal times.
- Inmates will be confined to the incident base or camp while off shift.
- Inmates shall not be allowed to handle explosives or detonating devices.
- Civilians and inmates shall have separate schedules for bathing.

[10] National Wildfire Coordination Center, *Wildland Fire Incident Field Management Guide: 2014* (Boise: National Wildfire Coordination Center, 2014) 120.

Smokejumpers Smokejumpers are elite firefighters who are highly trained and experienced. They arrive by parachutes to a fire. The *Fireline Handbook* clarifies the smokejumper resource as "a specifically trained and certified firefighter who travels to wildland fires by aircraft and parachutes to the fire."[11] The National Smokejumper Association also provides a summary of a smokejumper. "For more than 70 years, smokejumpers have enjoyed the admiration and appreciation of the American public, while performing a critically important service in protecting the nation's forests."[12]

Once the jumpers arrive on the fire, they are typically tasked with managing and suppressing the fire by themselves (depending on size, complexity, etc.). Smokejumpers maintain high fire qualifications and can typically fill important and critical fire management positions. In addition, they may be assigned for days to an incident without any logistical support. If logistics supplies are needed, they are often cargo dropped into the fire by airplanes. If a smokejumper crew does not arrive via an airplane, then they are considered a Type 1 Hand Crew.

According to the US Forest Service, there are over 270 smokejumpers employed by the US Forest Service, participating in fire activities.[13] They are assigned to Jumper Bases located in Idaho, California, Montana, Washington, and Oregon. The Bureau of Land Management also employees smokejumpers in Idaho and Alaska. Smokejumpers are also considered national resources and can be temporary stationed in areas away from their home unit during high fire activity.

Smokejumpers accessing a remote wildland fire.

Type 1 Helicopter dropping water on a fire.

Helitack Helitack crews work directly with helicopters to support their tactical or logistical activities. The crews are specially trained to work with helicopters. Some helitack crews have rappelling and/or short-haul capabilities. (Short-haul capabilities refer to a helicopter's ability to provide immediate medical assistance to an injured firefighter. They have qualified medical personnel and can winch or hoist firefighters out of difficult locations. They will transport (for a short distance/time) to a location for an approved medical helicopter to assume responsibility of the situation.)

---

[11] National Wildfire Coordinating Group, *Fireline Handbook*, 348.

[12] National Smokejumper Association *How Do I Become a Smokejumper*, accessed February 14, 2015, http://smokejumpers.com/index.php/becomeasmokejumper/get.

[13] United States Forest Service. Fire and Aviation Management, *Smokejumpers*, accessed February 14, 2014, http://www.fs.fed.us/fire/people/smokejumpers/.

The National Wildfire Coordinating Group provides an explanation of helitack as: "The utilization of helicopters to transport crews, equipment, and fire retardants or suppressants to the fireline during the initial stages of a fire. The term also refers to the crew that performs helicopter management and attack activities."[14]

Helicopters are classified into three (3) types. Type 1 Helicopters are the largest helicopters and are capable of transporting a higher level of equipment or firefighters. They can also provide the highest amount of water for suppression efforts. The information below provides a summary of the three types of helicopters used on a wildland fire.

TABLE 7-11 ***HELICOPTER MINIMUM SPECIFICATIONS***

Type of Helicopter	Allowable Payload (at 59° F at sea level)	Passenger Seats	Water or Retardant Capabilities (minimum)
Type 1	5,000	15 or more	700 gallons
Type 2	2,500	9 to 14	300 gallons
Type 3	1,200	4 to 8	100 gallons
Helitanker	N/A	No passengers	1,100 gallons

Courtesy of Brian Henington

Type 1 Helicopter.

Courtesy of Brian Henington

Type 2 Helicopter.

© Wollertz/Shutterstock.com

Type 3 Helicopter.

Courtesy of Brian Henington

Helitanker.

[14] National Wildfire Coordinating Group, *Glossary of Wildland Fire Terminology*, 99.

Dozers Fireline dozers (bulldozers) are used to construct fireline and support other fire activities. They are highly effective in line construction and are vital on large fires. Their speed and power decreases the number of firefighters needed to construct fireline. Often, hand crews will work in conjunction or behind dozers to reinforce the fireline. Dozers are limited in use due to certain topography features or rocky terrain. Dozers must have an enclosed compartment for the operator and should have lights for night operations. A dozer or **Heavy Equipment** boss is usually assigned with the dozer to facilitate supervision, ensure communications, and locate fireline placement. Firefighter Safety: Firefighters should not be close to dozer activities. The table below identifies the dozer type based on size and capabilities.

TABLE 7-12 *DOZER MINIMUM SPECIFICATIONS*

Type of Dozer	Category	Horse Power (Minimum)	Manufacture Examples	
Type I	Heavy	200	D8, D7, JD-950	© Anna Moskvina/ Shutterstock.com
Type II	Medium	100	D5N, D6N, JD-750	Courtesy of Brian Henington
Type III	Light	50	JD-450, JD-550, D3 or D4	Courtesy of Brian Henington

Heavy Equipment Heavy equipment is a relatively new category for wildland fire resources. This category includes any heavy machinery that can be used to support fire activities. Examples of heavy equipment include tractor/plows, road graders, track hoes, masticators, chippers, skidders, feller bunchers, hydro-ax, etc. Over the last several years, fire managers are becoming more resourceful in their utilization of fire resources. Utilizing specialized heavy equipment can improve suppression efforts while reducing the exposure to firefighters on the ground in high risk/hazard areas. An example may include the removal of snags with feller bunchers from a fireline in preparation of a large-scale burnout operation. Graders have demonstrated effectiveness on grass fires. Heavy equipment bosses will be assigned to work with individual equipment to ensure safety measures are in place and tactical objectives are prioritized and maintained. The photographs below illustrate some examples of heavy equipment used on wildland fires.

Courtesy of Brian Henington

Feller buncher removing trees along a fireline.

© vallefrias/Shutterstock.com

Hydro-ax constructing fireline.

© Dmitry Kalinovsky/Shutterstock.com

Road grader constructing fireline.

© Christian Delbert/Shutterstock.com

Track hoe feeding a wood chipper.

Firing Crews Fighting fire with fire is a proven and very effective suppression tactic used by fire managers across the United States. Not all crews have the qualifications or experience to conduct safe firing operations. Most firing operations will be conducted by Interagency Hotshot Crews or Type 1 Handcrews; however, certain fire agencies have internal crews that specialize in burnout operations or prescribed fire activities. These crews are managed by a Firing Boss. The images below provide an illustration of some examples of firing crews.

©James Bostian

Firefighter working as a member of a firing crew.

Courtesy of Brian Henington

Firing crew ignites fuel to achieve identified tactical objectives.

Courtesy of Brian Henington

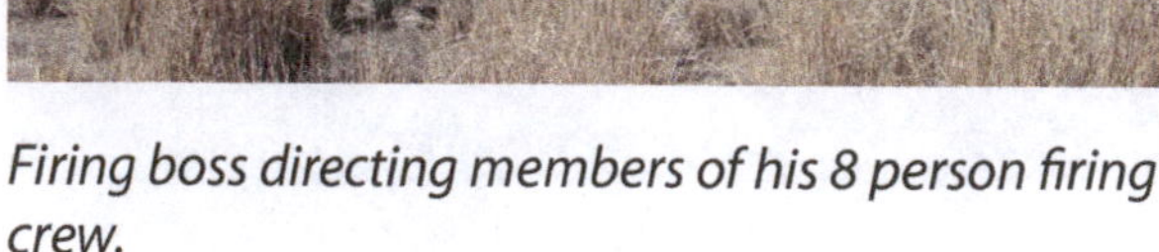

Firing boss directing members of his 8 person firing crew.

Courtesy of Brian Henington

Firing crew conducting nighttime burnout activities.

Felling Crews Felling crews specialize in cutting or falling trees. These crews are made up of **sawyers** or **fallers** (chainsaw operators) and **swampers** (individuals that support sawyers). These specialized crews can be agency firefighters or private contractors, such as professional loggers. Felling crews of professional loggers are very common in the Pacific Northwest, Northern California, Montana, and Idaho. The individual in charge of the felling crews is called a Felling Boss. To function as a sawyer, you have to meet agency specific or national standards for the size of tree being cut. There are three general falling standards or guidelines used by most fire agencies. The classification is based on the DBH (diameter at breast height) of the trees being cut. They are identified in the table below.

TABLE 7-13 ***FALLER OR SAWYER STANDARDS/QUALIFICATIONS***

Type of Faller/Sawyer	DBH Standards	Requirements Needed to Advance to Higher Level
A Faller	Up to 8 inches	Must have NWCG S-212 before operating a chainsaw. NWCG S-212*
B Faller	Up to 24 inches	A Faller (efficient) and certification process
C Faller	Unlimited (can also be based on hazard rating or DBH specification)	B Faller (efficient) and rigorous certification process

**NWCG S-212 Wildland Fire Chainsaws*

Courtesy of Brian Henington

Wildland Firefighter Mason Bell "gunning" his face cut on a C tree.

Courtesy of Brian Henington

Sawyer preparing to drop a B tree. He is being supported by his swamper.

Courtesy of Brian Henington

Sawyer dropping a hazard tree (snag).

Courtesy of Brian Henington

A sawyer working a fallen hazard tree that is wedged between other trees. A tree like the one identified here increases the risk and exposure to the firefighter. Only qualified sawyers and sawyers who are comfortable with proceeding should undertake this activity.

Water Tenders Water tenders are essential for supplying water to areas without available water. They are used extensively in certain parts of the country because adequate water sources are not available. Firefighting efforts are greatly improved when water is used. They can also be used to mitigate dust concerns along roads or at helispots. The new trend is tactical water tenders. The tactical water tender can provide both water support or can be involved in suppression efforts. Unlike a normal water tender, the tactical tender must have a qualified engine boss associated with it due to the tactical consideration. The table below provides a classification of water and tactical tenders.

TABLE 7-14 ***WATER TENDER STANDARDS/QUALIFICATIONS***

Type of Tender	Tank Capacity (minimum)	# of Firefighters (minimum)	Pump and Roll Capabilities
S1 (Support)	4,000 gallons	1	No
S2 (Support)	2,500 gallons	1	No
S3 (Support)	1,000 gallons	1	No
T1 (Tactical)	2,000 gallons	2	Yes
T2 (Tactical)	1,000 gallons	2	Yes

**NWCG S-212 Wildland Fire Chainsaws*

Courtesy of Brian Henington

Support water tender.

©James Bostian

Tactical water tender holding fireline during burnout operations.

Air Tankers "Fixed-wing aircraft certified by FAA as being capable of transport and delivery of fire retardant solutions."[15] Working with air tankers requires the ability to identify concerns and being able to communicate the locations of those concerns. Coordination with Air Tankers will usually involve highly qualified firefighters. When in the air, air tankers are managed by the Air Tanker/Fixed-Wing Coordinator or the Air Tactical Group Supervisor. Air tankers can deliver **retardant** (slurry) or water to areas of concern. The misconception by the general public is that air tankers extinguish fires. They use retardant or water that is intended to slow a fire down so crews on the ground can reinforce their actions. Working under air tankers can be dangerous to firefighters (this concept will be discussed in later chapters).

WHY DO WE USE FIRE RETARDANT?

"Today, retardant is aerially applied and is known to reduce the spread and intensity of fires and slow larger, more damaging, and thus, more costly fires. In many situations, using retardant to fight fires is the most effective and efficient method of assisting firefighters in protecting people, resources, private property, and facilities; sometimes it is the only tool available." (US Forest Service)

As is the case of all resources identified in the section, air tankers are also classified on their capabilities. Type 1 Air Tankers are the largest of the four types and can deliver the most retardant or water to a fire. The newest air tanker on the scene is the Very Large Air Tanker (VLAT). The VLAT is a DC-10 airplane that carries 11,600 gallons of fire retardant.

TABLE 7-15 ***AIR TANKER STANDARDS/QUALIFICATIONS***

Type of Air Tanker	Minimum Capacity	Examples
Type 1	3,000 gallons	© TFoxFoto/Shutterstock.com
Type 2	1,800–2,999 gallons	Tanker 45. Type 2 Air taker. Courtesy of Brian Henington

(continued)

[15] Ibid., 26.

Type 3	800–1,799 gallons	 © Heather Lucia Snow/ Shutterstock.com
Type 4	Less than 800	Single Engine Air Tanker (SEAT).  © David Molina Grande/ Shutterstock.com

TRANSPORTATION SAFETY AND CONCERNS

Transportation safety can be best described as the mode of transportation used to get firefighters to their destination and to return them home after their mission. The risk with transportation cannot be understated. The most common mode of transportation used on fires is your assigned vehicle. This may be a support vehicle, an engine, a crew carrier, or a water tender. Vehicle accidents involving firefighters is high, not low. We often travel across states to reach our destination. The longer we are on the road, the greater the risk of an accident occurring. This section describes our main modes of transportation and the safety measures we should have in place to ensure we arrive and return safely from assignment.

Courtesy of Brian Henington

Strike Team of Type 6 Engines.

Driving Safety

The National Interagency Fire Center (NIFC) is responsible for managing and tracking wildland firefighter line-of-duty injuries and fatalities. NIFC categorizes injuries and fatalities based on type of accident, year, and location. The data is very useful for firefighters to research and understand the activities that are at high or elevated risk. Driving is one of those risks. One hundred and thirty-nine driving accidents have occurred since 1910. Out of those accidents, 110 have resulted in fatalities. The accidents have occurred due to vehicle rollovers, or being hit by an engine, or another type of vehicle accidents. Speed, driver inattention, or mechanical failures have all been associated with these accidents. The figure below depicts the driving accidents that have involved firefighters. This should illustrate how serious we must approach driving our fire vehicle.

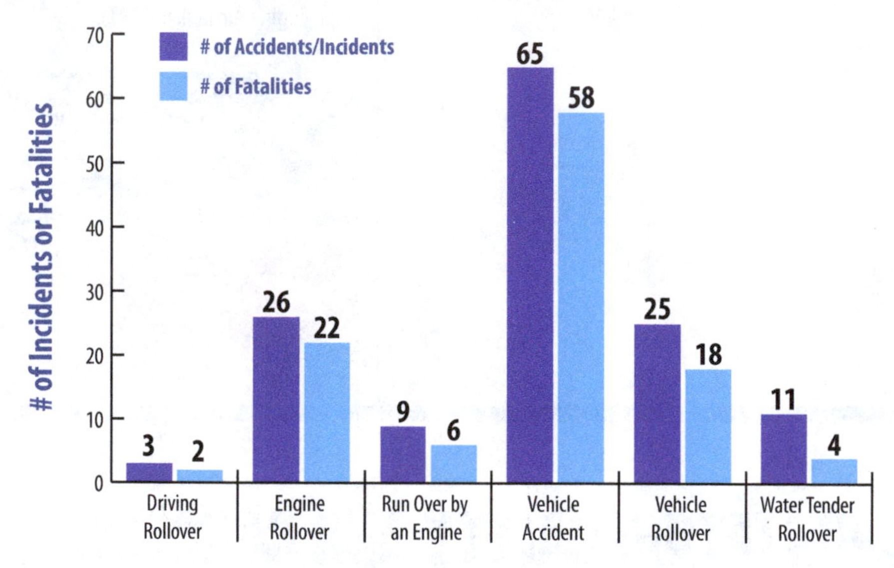

Figure 7-2 *Vehicle Accidents and Fatalities by Type (Involving Firefighters)*

Driving Standards After a long and hard day of work, firefighters are exhausted. Driving home from an incident may be more of a hazard or risk for tired firefighters. You should monitor your driver's condition and fatigue levels. It may be better to bed down and travel in the morning than risking driving home after an exhausting day. In addition, fire managers have now instituted standards for driving requirements. The standards involve both driving long distances to fire incidents or initial attack activities. These rules are outlined in the *Wildland Fire Incident Management Field Guide—2014 Edition.* Below is a brief summary of the driving standards outlined in the manual.

- Follow the work to rest ratio. Clarification. You began your workday at 0700 hours. At 1700 hours your engine receives a dispatch to a new fire. The fire requires 2.5 hours of drive time before you reach the incident. You have already worked 9 hours (excluding lunch). The fire is declared controlled at 2130 hours. You have worked 14 hours and know the drive home will require 2.5 more hours—putting you at 16.5 hours worked in a day. This exceeds the work to rest ratio. The fact of the matter is that if you get in an accident, it will be identified that you exceeded the work to rest ratio. It may be better to bed down (for 8 hours), then return to your fire station the next day. It's all about managing risk.
- A driver can only drive for 10 hours a day. That does not mean a vehicle cannot operate more than 10 hours. The vehicle can continue activities if other qualified drivers are available.

2200 Hundred Hour Rule Although this rule is not covered or identified in the *Wildland Fire Incident Field Management Guide*, many agencies are do not allow driving or operating a vehicle after 2200 hours (10:00 pm), while driving to incidents that are away from your home unit. This does not necessarily apply to initial attack fires. For example, if you are traveling from Texas to support a fire in Nevada, the vehicle must stop and firefighters bed down at 2200 hours. This can occur at a truck stop or a state park. Good supervisors will plan for this rule and locate a motel before the 2200 hours to provide their firefighters with adequate rest and shelter.

Driver Considerations

- Drivers should be certified by agency standards to operate their assigned vehicles. Some fire vehicles require a Commercial Driver's License. Other agencies require emergency driving endorsement or defensive driving. Drivers should adhere to the standards that manage their driving activities.
- Emergency (Code 3) Driving: Agency specific. Most wildland agencies only use their overhead emergency lights during the initial stages of an incident or during nighttime activities. Do not touch emergency lights until you have verified the standards established by you agency.
- Do not talk or text on cell phones.
- Drivers should leave radio communication to the co-pilot.
- Maintain situational awareness and drive defensively.
- Co-pilots are responsible for navigation and ensuring the driver is alert and focused on driving.

Safety Measures for Vehicle Transportation

- Adhere to the standards of your fire agency.
- Keep tools secured in approved boxes or cages. Tools cannot be unsecured inside a vehicle with firefighters. Cutting edges need to be protected.
- Wear seat belts even in the back of a crew carrier or on a bus.
- No smoking in a government vehicle.

Boat Transportation

Often, firefighters will be transported to specific locations of the fire by boats. Boats will belong to a fire or government agency or could be contracted with private parties. Ensure you respect the boat and the owner.

Safety Measures for Boat Transportation

- Adhere to the standards of your fire agency.
- Keep tools secure and cover cutting edges.
- Wear life jackets.
- Remain seated.
- Avoid horseplay.

Courtesy of Brian Henington

Helicopter travel is a very common mode of transportation on wildland fires.

Helicopters

It may be necessary to transport firefighters to remote locations on the fire by helicopters. Helitack will be responsible for ensuring you follow the standards and rules associated with flying on a helicopter. Helitack will also be responsible for loading and unloading all equipment/tools. Helitack will also weigh you using a weight scale. Helicopters can only carry a specific amount of weight depending on altitude. If the helicopter is carrying 10 firefighters who have all under-exaggerated their weight by 20 pounds, then the helicopter would be carrying an additional 200 pounds. This may cause an issue with the helicopter; thus the importance of being weighed prior to flying. Helpful hint, the fire gear on your body will increase your total body weight by 10 to 15 pounds.

Some people have a fear of flying on a helicopter. If you are one, let your supervisor know of your concerns.

Helpful Guidelines When Preparing to Fly on a Helicopter

- Helitack will load tools and equipment. Your responsibility is to ensure the cutting edges of your tools are covered.
- Do a PPE inspection on yourself and your crew members. Is your chinstrap on and your helmet snug and secure to your head? Button your top button on your fire shirt. Ensure everybody's sleeves are rolled down and fastened or buttoned.
- Is your eye protection on and secure to your face? Is your ear protection in?
- Ensure you are weighed or give a very accurate weight to helitack.
- Receive a briefing from helitack.
- Follow the instructions of helitack.
- Pay close attention to the emergency landing protocol.

Approaching a Helicopter

- Only approach from the front of the helicopter after you have been told to do so.
- Never approach a helicopter from the rear. It is extremely dangerous.
- Do not approach the helicopter if you are higher than the helicopter. For example, the helicopter has landed in a depression and you are above it waiting to load.
- Walk in a low or a crouched position toward the helicopter.

- Stay calm. This may be your first helicopter ride and it can be exciting to some or scary to others. Relax! Listen to the rules of the helitack and follow their instructions. It may be appear safe to approach a helicopter, but wait until helitack tells you to or the pilot has motioned to you to approach. Before you approach, keep your eyes on the pilot. This technique will ensure you are gauging the pilot's level of comfort or concern.

Helpful Techniques When Flying on a Helicopter

- Hardhat should be secured to your head with the supplied chinstrap.
- No smoking or chewing tobacco.
- Tools and fire packs will be handled by helitack.
- Fasten your seat belt. The first time you put on a helicopter seat belt it may be confusing. Allow helitack to assist you. If you do not have your seat belt fastened, let somebody know.
- Helitack will close and secure doors. If the helicopter does not have doors, make sure your hands and body parts are inside the helicopter at all times.
- Do not distract the pilot, especially during landing or take off.

Leaving or Unloading from a Helicopter

- Follow the instructions on when to unload. Listen to helitack or the pilot. Do not unload until they have told you to do so.
- Walk in a low or crouched position away from the helicopter.
- Do not leave uphill from the helicopter.

The most dangerous consideration related to helicopters is the rotors. You have to stress and remember the importance of crouching as you approach and leave a helicopter; never approach from the rear or uphill of a helicopter. Rotor wash is also a hazard. This is why we secure our hard hat to our heads with a chinstrap.

Allow helitack to do their jobs! Follow their instructions at all times!

Commercial Airplane

Traveling from one location of the country to another usually involves commercial airlines. You should check with your local or zone dispatch for a list of fire-related items you can or cannot carry on a commercial jet. You will have to go through the same screening process as civilians.

A Few Commercial Aircraft Recommendations

- Multi tool and/or any knife should be in your checked baggage—not on your person.
- Do not carrying any ignition or firing devices in your baggage, including your fire pack.
- Tools and chainsaws should be checked out when you arrive at your assigned fire. Do not take your favorite tool with you. Check one out when you arrive at the fire.
- Many airlines limit you to two bags. Fees for extra bags may or may not be reimbursed by your fire agency. Make sure you check with your agency before you fly.
- Adhere to 65-pound maximum limit on your fireline gear and personal bag.

Foot Travel

> **FOOT TRAVEL KEY SAFETY COMPONENT**
>
> Maintain a minimum of 10 feet between each firefighter while walking to or from your destination.

To reduce risk of a tool from another firefighter hitting you, ensure you maintain the following rule of thumb: minimum of 10 feet between each firefighter. A crew of 20 firefighters (from the first firefighter to the last) should be at least 200 feet long or 67 yards. This standard will allow us to manage the risk better.

Examples of Common Foot Travel Hazards Include:

- Tree Related (most of these hazards will be explained in detail in further chapters).
  - Snags
  - Leaning trees
  - Whipping branches
  - Widow-makers
  - Stump holes (these can hold heat and have been associated with firefighter burns)
- Environmental
  - Darkness
  - River or stream crossings
  - Rolling rocks
  - Rolling logs
  - Uneven terrain
  - Rocks or boulders
  - Snakes
  - Poisonous insects
  - Poisonous plants
- Human or Firefighter Related
  - Sharp hand tools
  - Other firefighters tools
  - ATVs
  - Vehicles
  - Heavy equipment
  - Aircraft

Courtesy of Brian Henington

The New Mexico Veterans (Returning Heroes) Crew walking to their assigned location on hazardous volcanic rock.

SUMMARY

Firefighting resources are the most important asset to firefighter operations. They work together to contain and control a fire. Without these resources, fires cannot be suppressed. You may be hired to work on a hand crew or engine crew that will require a specific set of skills and knowledge to be effective and safe. As you progress throughout your career, you may move to higher skilled resource such as helitack, hotshots, or smokejumpers. The key to all of these resources is that they perform their assigned duties in the safest manner possible.

Transportation safety is also a concern to us. Understand the hazards associated with each type of travel mode, and apply those risk management techniques to ensure you arrive and return home as safely as possible.

KNOWLEDGE ASSESSMENT

1. What is a resource type and kind? Provide an example of each.
2. What are four types of resources used on a wildland fire?
3. What is a single resource?
4. Explain the difference between a task force and a strike team.
5. How many resources are involved in a strike team or task force? What is the ideal ratio of supervisor to subordinate/resource?
6. What is the major difference between Interagency Hotshot Crews and other hand crews?
7. How do smokejumpers access a fire?
8. What engine is considered the workhorse on wildland fires?
9. What is the biggest threat or risk related to a helicopter?
10. How many accidents and fatalities have been associated with driving activities on a fire?
11. A driver of fire equipment cannot operate the apparatus for no more than ____________ hours on any given day.
12. Should a driver of fire equipment use a cell phone or radio? If not, who should?
13. What are some hazards associated with foot travel?
14. What is the minimum distance firefighters must maintain between each other when walking to their assigned work area?

EXERCISES

1. Research the requirements and job responsibilities of a smokejumper. You can conduct this research using outside references, such as searching the Internet.
2. Explain the qualifications of chainsaw operators. What is the basic level of training required by the National Wildfire Coordinating Group? How does a sawyer move from one category to the next?
3. Conduct an Internet search of a vehicle accident involving a firefighter (either wildland or structural). How did the accident happen? Where was it located? Did you find any that occurred near or where you live? What would be your recommendation to ensure a similar accident does not occur in the future?

BIBLIOGRAPHY

Federal Emergency Management Agency. *IS-703.A: NIMS Resource Management*. United States Department of Homeland Security, 2010.

National Interagency Fire Center. *Historical Wildland Firefighter Fatality Report*. Boise: National Interagency Fire Center. Boise. http://www.nifc.gov/safety/safety_HistFatality_report.html.

National Wildfire Coordinating Group. *Fireline Handbook*. Boise: National Wildfire Coordinating Group, 2004.

National Wildfire Coordinating Group. *Glossary of Wildland Fire Terminology*. Boise: National Wildfire Coordinating Group, 2014.

National Wildfire Coordination Center. *Wildland Fire Incident Field Management Guide: 2014*. Boise: National Wildfire Coordination Center, 2014.

National Smokejumper Association. *How Do I Become a Smokejumper*. http://smokejumpers.com/index.php/becomeasmokejumper/get.

United States Forest Service. "Aerial Application of Fire Retardant" in *Seeing Red: A Short History of Fire Retardant and the Forest Service*. United States Forest Service. http://www.fs.fed.us/fire/retardant/history.html.

United States Forest Service. National Interagency Hotshot Crews (IHC). *United States Forest Service*. http://www.fs.fed.us/fire/people/hotshots/IHC_index.html.

United States Forest Service. Fire and Aviation Management. *Smokejumpers*. http://www.fs.fed.us/fire/people/smokejumpers/.

CHAPTER 8

THE RULES OF ENGAGEMENT

Learning Outcomes

- Analyze common risk management terminology used on wildland fire suppression activities.
- Describe the importance of LCES and how this risk management concept should be used by wildland firefighters on every fire.
- Express the importance of communication and how it relates to safe firefighting.
- Evaluate the importance of safety zones and escape routes.
- Recognize the common denominators of wildland fire fatalities.
- Explain the importance of the Standard Firefighting Orders and how they should be used at all times during suppression activities.
- Describe the Watch Out Situations and how firefighters should implement them during fire activities to ensure safety is prioritized and maintained.

Courtesy of Brian Henington

Key Terms

Burnover
Burn Period
Communications
Deployment Site
Entrapment
Entrapment Avoidance
Escape Routes
LCES
Lookout
Mitigation Actions
Near Miss
Risk Management
Rules of Engagement
Safety Zones
Standard Firefighting Orders
Tactical Pause
Watch Out Situations

OVERVIEW

Wildland firefighters utilize rules and standards that enable them to determine if a fire can safely be engaged. These national standards and rules are considered the wildland firefighter **Rules of Engagement**. The term "rules of engagement" was adopted from the military and law enforcement arena. These two professions use rules that dictate how they are to engage a hostile or threatening individual or individuals. Wildland fire managers have developed similar rules with the same intent and purpose. If a fire has the potential to harm firefighters then it should be viewed as a hostile or threatening living organism.

The purpose of this chapter is to explore our Rules of Engagement and to provide a short analysis of each component. You will become familiar with the concept of LCES; the Standard Firefighting Orders; The Common Denominators for Wildland Firefighter Fatalities, and the Watch Out Situations. Together these rules and guidelines function as a tool to establish the safest tactical solution for firefighters. These rules must be in place prior to exposing firefighters to the risks associated with fire suppression.

KEY CONCEPTS

The risk management techniques and standards used in wildland firefighting are the most important facets of safe firefighting. These concepts were designed to provide a safe atmosphere in a hostile environment. For the purpose of this chapter, we have provided some key definitions of important risk management concepts. As a firefighter, you need to familiarize yourself with these concepts and ensure you understand their intent and purpose.

- **Burnover**: An event in which a fire moves through a location or overtakes personnel or equipment where there is no opportunity to utilize escape routes and safety zones, often resulting in personal injury or equipment damage.[1]
- **Burn Period**: Identified as the time period between 1400 hours to 1700 hours. Illustrates the most dangerous part of the day for extreme or elevated fire behavior.
- **Entrapment**: A situation where personnel are unexpectedly caught in a fire; behavior-related, life-threatening position where planned escape routes or safety zones are absent, inadequate, or compromised. An entrapment may or may not include deployment of a fire shelter for its intended purpose. These situations may or may not result in injury. They include "near misses."[2]
- **Entrapment Avoidance**: A process used to improve the safety of personnel on the fireline, which emphasizes tools and tactics available to prevent being trapped in a burnover situation. This process includes appropriate decision making through risk management, application of LCES, use of pre-established trigger points, and recognition of suitable escape routes and safety zones.[3]
- **Mitigation Actions**: Actions that are implemented to reduce or eliminate (mitigate) risks to persons, property, or natural resources. These actions can include mechanical and physical tasks, specific fire applications, and limited suppression actions.[4]

---

[1] National Wildfire Coordinating Group, *Glossary of Wildland Fire Terminology* (Boise: National Wildfire Coordinating Group, 2014), 40.

[2] Ibid., 66.

[3] Ibid.

[4] Ibid., 122.

- **Near Misses**: Any potential accident which, through prevention, education, hazard reduction, or luck, did not occur.[5]
- **Risk Management**: A continuous, five-step process that provides a systematic method for identifying and managing the risks associated with any operation.[6]
- **Rules of Engagement**: The rules or standards wildland firefighters use as the guiding instrument for engaging a wildland fire.
- **Tactical Pause**: Disengaging the fire for a short period to evaluate tactical activities and reevaluate safety considerations and the Rules of Engagement. A tactical pause should be taken in a safe location. Are the activities you are engaged in now considering what the fire behavior will be during the burn period? Do your current activities make sense? Are they safe?

LOOKOUTS, COMMUNICATION, ESCAPE ROUTES & SAFETY ZONES (LCES)

One of the most effective risk management concepts used in wildland firefighting is LCES. This concept stands for **Lookouts, Communications, Escape Routes**, and **Safety Zones (LCES)**. LCES should be used in conjunction with the **Standard Firefighting Orders** as part of the Rules of Engagement for firefighters. The process is not only used during the early stages of a fire, but it should constantly be used and reevaluated throughout the day and operational period. Once LCES becomes compromised, then actions (in theory) should cease until the compromised component is reestablished. "LCES must always be in place when working on the fireline. It is a tool that looks at the **Watch Out Situations** to evaluate the assignment, identify the hazards, analyze the risks, and implement steps to ensure safety by interconnecting with the Standard Firefighting Orders, as its operational component."[7]

Figure 8-1

LCES was created by a legend in the fire service. Paul Gleason was a hotshot superintendent, career wildland firefighter, and fire ecologist. After he retired from the fire service, he continued his passion for wildland fire safety by teaching aspiring firefighters as a professor at Colorado State University. Professor Gleason created LCES to be used as a simple and functioning tool to determine if the actions and mitigation measures needed to suppress a fire are adequate and warranted.

---

[5] Ibid., 127.

[6] Ibid., 151.

[7] National Wildfire Coordinating Group, *Firefighter Training, S-130* (Boise: National Wildfire Coordinating Group, 2003), 4B.2.

THE IMPORTANCE OF LCES

The Fire Science Program at Central New Mexico Community College preaches and teaches the importance of LCES. We do not teach LCES as a replacement for the Standard Firefighting Orders; we teach it as a tool used in conjunction with our primary fire doctrine—Standard Firefighting Orders. Twenty years of teaching thousands of firefighters has taught me one important thing about the learning comprehension of entry-level firefighters. It is easier for a new firefighter to understand the concept of LCES than to memorize the Standard Firefighting Orders and Watch Out Situations.

As you know, the common approach for fire supervisors is to demand that their firefighters memorize the Standard Firefighting Orders. They usually do so with threats. Learn and recite these rules within two weeks or pay the consequences! This philosophy is not conducive to learning. This method of learning does one thing: It forces firefighters to memorize the rules, not to comprehend and understand why we even have these rules. We used to incorporate this same philosophy. The students were able to recite the rules word-for-word at the end of the semester. When the next semester started, I would ask them to recite the rules again. They would not be able to. I then would ask them to clarify what LCES was. They could without hesitation.

Remember, Professor Gleason created these rules to facilitate the understanding and comprehension of the Rules of Engagement. When a firefighter truly understands LCES, they then will begin to understand how these concepts directly relate to the Standard Firefighting Orders and Watch Out Situations and not only be able to recite them, but understand their purpose.

The creation of LCES was one vested in history; a fatality fire that claimed the lives of six firefighters. Paul Gleason was the superintendent of the Zig Zag Hotshots. In 1990, he was assigned to the Dude Fire near Payson, Arizona. The Dude Fire started as a lightning fire on June 25. Within an hour of detection, the fire had demonstrated extreme burning indices and had grown to 50 acres. Because of the size of the fire and the threat to a subdivision, the fire became a top priority with fire mangers. They requested a Type 1 Incident Management Team and 18 hand crews (360 firefighters) to assist their suppression efforts.

One of the crews assigned to the fire was the Perryville Type 2 Hand Crew. The 20-person inmate hand crew arrived on the fire on the evening of June 25 and soon began firefighting activities. On June 26, they were assigned, with additional crews, (including the Zig Zag Hotshots) to begin efforts to protect the subdivision. The fire behavior throughout the morning of June 26 was minimal. As the firefighting efforts continued through the day, fire behavior began to increase. The column of the fire began to collapse, causing the fire to transition from a surface fire to a crown fire. Experts believe the fire transitioned to a crown fire at approximately 1420 hours.

The Perryville Hand Crew tried to escape, but was caught off by the advancing fire. They made the decision to deploy their fire shelters. The crew representative, Dave LaTour, advised the fire managers via radio that the crew was deploying fire shelters. Prior to the blowup, the Zig Zag Hotshots and other hotshot crews had taken refuge in a predetermined and adequate safety zone. They made the determination to disengage the fire until the behavior decreased.

Paul Gleason had overheard the message of the fire shelter deployment. He and two other hotshot superintendents risked their lives and headed to the area of the deployment. When he arrived on scene, he saw the bodies of six firefighters. The Dude Fire fatalities were devastating. It was overwhelming. After this incident, Paul studied the events of this tragedy. He wanted to develop a tool for fire fighters to use to avoid situations similar to this. After a winter of reflection and consultation with other highly experienced firefighters, Paul devised this tool to be used by firefighters as a supplement to the Rules of Engagement because he felt the other rules used by firefighters were too complex or not being used appropriately.

Author Note: About a year ago, I was instructing an engine boss class at a prestigious wildland fire academy. The class had very skilled and experienced firefighters—hotshots, helitack, and seasoned engine crew members. I asked them to tell me the history of LCES. Not one person knew. They all knew the essentials and importance of LCES, but they were unaware of why they were created. I shared the story, the reason Paul Gleason created this concept, and what event led to the creation. Paul Gleason was tired by the tragedies he had witnessed in his career. He wanted to create an easy tool to be used and implemented by all fire personnel, not just fire managers. He wanted to create a concept that all firefighters can understand and comprehend.

PAUL GLEASON—CAREER FIREFIGHTER

Paul Gleason's career as a firefighter spanned parts of five decades, starting as an 18 year-old crew member on a Southern California hotshot crew and culminating as a college professor of wildland fire science.

Paul grew up in Southern California, the son of a traveling evangelist preacher. He became an accomplished rock climber in his teens and continued to climb through his entire life. In 1964 he got his first job as a firefighter on the Angeles National Forest. He continued to work there on the Dalton Hotshot Crew through 1970, with the exception of a one-year stint in the U.S. Army. From 1971 to 1973 he went to college and earned a degree in Mathematics.

During this time he also traveled and climbed extensively. He returned to work as a firefighter in 1974 as the Assistant Foreman for a 20-person Regional Reinforcement Crew on the Okanogan National Forest. Then in 1977 he took the job as the Assistant Superintendent of the Zig Zag Hotshot Crew on the Mount Hood National Forest, moving up to Superintendent in 1979. He remained in that role until 1992. He then transferred to the Arapaho-Roosevelt National Forest as a District Fire Management Officer and eventually became the Forest Fire Ecologist. His next move was to another fire agency in 1999 as the Deputy Fire Management Officer for the Rocky Mountain Region of the National Park Service. Mandatory retirement at age 55 took Paul away from the federal fire service in 2001 and into academia. For the next two years Paul was adjunct professor for the Wildland Fire Science program at Colorado State University. He remained in this role until he lost his battle with cancer in 2003.

During his career Paul Gleason was front and center on three significant fires of the modern era—the Loop Fire in 1966, the Dude Fire in 1990, and the Cerro Grande Fire in 2000. His role on these three touchstone fires gave rise to his passion for firefighter safety and the "student of fire" philosophy that he crusaded for. He was a leader of firefighters and he was a leader for the wildland fire service. Paul's contributions are far reaching. He teamed up with D. Douglas Dent and pioneered the professional tree falling program for wildland firefighters. He developed the LCES (1991 document on LCES by Gleason) concept that has become the modern foundation of firefighter safety. He was very involved in the development of fire behavior training... with a focus on taking the scientific aspects of extreme fire behavior and making them understandable concepts for every firefighter. He reached outside the fire service and collaborated with experts, such as Dr. Karl Weick, who were doing research in the realm of decision-making and high reliability organizations.

In the final tally, as always, Paul was a role model "student of fire." To the very end of his life he was engaged in teaching and learning about fire.

Wildland Fire Leadership Development Program, *Leaders We Would Like to Meet—Paul Gleason*, accessed February 22, 2015.

How LCES Works

LCES is an essential element of our Rules of Engagement. "The nature of wildfire suppression dictates continuous evaluation of LCES and the constant reevaluation and establishment of LCES as the fire grows."[8] If one of the elements is compromised or violated, then actions should cease until they are reestablished. A **lookout** is a person who can see the fire and the crew at all times. He or she is your extra eyes. Lookouts are critical to safe firefighting. If the crew's lookout has to abandon his or her location because the fire is threatening their location, then LCES is compromised. Stop actions until the lookout is reestablished! If a lookout is not present or adequate, then do not engage the fire until a lookout is established.

Communications is critical to our safety and efficient firefighting efforts. Not only is it used to ensure instructions are given and understood, but it is vital to safety and risk management. If you arrive on a fire without adequate communications with other resources or the incident commander, do not engage the fire until communications is established. If communications is lost during the day, then take a tactical pause and address the concern.

Escape routes can be best described as the path or route that takes you to safety or your safety zone. The National Wildfire Coordinating Group elaborates on escape routes as: "A preplanned and understood route firefighters take to move to a safety zone or other low-risk area. When escape routes deviate from a defined physical path, they should be clearly marked (flagged)."[9] There should be a minimum of two adequate escape routes at all times. If you do not have at least two escape routes, then do not engage the fire until you do.

The final component of LCES is safety zones. A **safety zone** is an area big enough for you, your crew, other crews, and equipment to escape to if fire behavior dictates it. Safety zones can be natural, manmade, or the fire itself. A safety zone is big enough so that you or your crew would not have to deploy a fire shelter to protect yourself from the fire. If you have inadequate safety zones, do not engage the fire until you have mitigated the concerns and addressed them.

As an entry-level firefighter, you will not be the primary person or persons responsible for implementing LCES; however, you must know that LCES should be in place at all times and that the measures are appropriate. If they are not, then your individual safety can be compromised. Ask yourself the following: Do we have a lookout in place and who is it? Is communications established and has the boss told us the backup plan if our primary method fails? Do I know where my escape routes are and are they identified? Is the safety zone close enough for me to get to and is it big enough that I would not have to deploy a fire shelter?

Can you now see how important it is that you understand and comprehend the essentials of LCES? If you feel like the LCES actions being used on your crew or fire are inadequate, then voice your concerns.

Most of the time, LCES can be implemented by a seasoned firefighter in a short time. Firefighters should be considering the concepts before actions occur. Once they have the measures in place and have discussed how to ensure the four components are maintained, then engage the fire. The information below describes in detail the four components of LCES. Always remember why Paul Gleason created these standards and how essential they are to your personal safety.

[8] Ibid., 4B.3.

[9] National Wildfire Coordinating Group, *Glossary*, 67.

Lookouts

Lookouts should be posted at all times. Rule # 5 of the Standard Firefighting Orders specifies that lookouts should be posted when there is possible danger. All fires are potentially dangerous; therefore, post lookouts at all times. Lookouts should be in a location where they have a good vantage point of the fire and their crew. They must be able to see the fire and their crew at all times. If they cannot, they need to inform the crew and move until they can effectively see both the crew and the fire.

Paul Gleason encouraged that a lookout should be an experienced, competent, and a trusted firefighter. In fact, he recommended that a lookout on a hotshot crew should be the hotshot superintendent. His view was that an experienced and highly-trained firefighter would be able to notice changes in weather, fire behavior, and activity quicker than an inexperienced firefighter would be able to.

The lookout must know the location of their crew's safety zones and where the escape routes are. If fire activity has the potential to compromise either one of these, he or she must alert the crew immediately. Another concern is when a crew is scattered over a long area of the fireline. This may require more than one lookout. The main factor that will dictate the need for additional lookouts is when the primary lookout can no longer see all the members of the crew.

The lookout will take weather readings on the hour and monitor trends. Is the temperature increasing? Is relative humidity decreasing? Are winds increasing? Do you notice thunderstorm development? All of these factors can contribute to increased fire behavior and additional exposure to the firefighter.

On large fires, a lookout may be established on an adjacent ridge. Make sure everybody knows the location. A signal mirror is a good tool to identify the location. On smaller fires, the incident commander or engine boss can function as a lookout. They should not be engaged in tactical activities, but remain mobile and alert to ensure they are providing an adequate lookout. You can also utilize air resources as a temporary lookout, but remember they have other duties. Do not rely on aircraft to provide adequate lookouts for you or your crew.

Courtesy of Brian Henington

Finally, another job duty of a lookout can be a relay for communications. Lookouts are usually in high locations and are able to function as a human relay for radio traffic. Lookouts should be busy, even though they may be stationary. They should be constantly evaluating fire behavior, weather, tactics, and communicating on the radio. This is not a job for one to take lightly or considered a break from tactical activities. That is why it is so important to have a lookout who is trusted and experienced and a great communicator.

Recommended Lookout Kit:

- Map
- Weather Kit
- GPS Unit
- Signal Mirror
- Radio with extra batteries
- Incident Action Plan
- Cell phone
- Binoculars
- Compass
- Food/Water
- Watch with stop watch

Communications

The ability to communicate on a fireline is essential. It has to occur! From the beginning of firefighting efforts, communication, or the lack of, has been at the center of far too many fatalities and accidents. We have to be willing to effectively communicate (which includes listening) to all fire personnel, not just our own crew.

On the fireline, the first thing that should be established is common radio frequencies. Do not engage a fire until you can talk to other resources. Program your radio and test the frequencies to ensure they are functional. Discuss a backup plan. What do we do if the radio is not working? Unfortunately, radio issues are common. Make sure you have a backup plan in place prior to engaging a fire. Do not count on your cell phone to provide adequate communications. Most of the fires we fight do not have adequate cell phone coverage.

Courtesy of Shauna Henington

Communication should occur whenever a change is occurring. Ensure you relay all pertinent information. Examples of changes include: fire behavior, weather, reassignment of crews to new tasks, etc. Do not assume that other people are aware or have heard the changes. Make sure they are informed. A new and proven standard is to confirm the message you are receiving. An example of this is: a weather reading is given at 1100 hours. The lookout reads the weather readings over the radio. All resources working with the lookout should confirm that they have received the message. If one resource does not, then the lookout should continue asking for confirmation until confirmation is given. This method verifies that all resources are hearing and comprehending the intended message.

Radios are essential to communications, but the preferred method of communications is face-to-face. Face-to-face communications allows for immediate information transfer and allows for clarification if the message is not understood. Examples of face-to-face communications include briefings (morning, initial supervisor briefing, tactical briefing), after action reviews, and direct tactical assignments. The information below illustrates some key points related to briefings:

- Briefings should be to the point, informative, and allow for questions.
- Briefings are a key component of effective and safe fire operations. If you do not receive a briefing before you engage a fire, do not proceed until you get the briefing. Demand that you receive a good briefing.
- Consider face-to-face briefings when situations change.

The most important component of effective communications is listening. You cannot effectively communicate if you do not listen. Have you ever had a boss who never listened to you? It's very frustrating, right? Ensure you listen to messages you are receiving and write them down. During chaotic times of a fire, you will not be able to remember all the information. Write the messages down. Learn to use techniques to improve your listening skills. The information below helps with listening skills.

- Focus on the individual communicating. Allow the individual to finish before you ask questions.
- Be respectful. Pay attention and make eye contact.
- Pay particular attention to non-verbal communication.
- Ask questions if you do not understand.
- Repeat the information to ensure everybody is on the right page.

Escape Routes

Escape routes are the paths that lead you to safety. According the Wildland Fire Lessons Learned Center, "Ninety percent of the time, the fireline is the escape route."[10] They are essential to safe firefighting activities. "Escape routes are probably the most elusive component of LCES. Their effectiveness changes continuously."[11] Escape routes should constantly be evaluated as the crew's progress with their suppression activities. Do not assume the escape route you used this morning will be adequate during the burn period. Timing is essential. If you feel the fire behavior has increased to a point that will jeopardize your safety, then consider withdrawing from the area.

You must always have at least two (2) escape routes. Escape routes must be free of debris such as rocks and logs. If you have to escape, you want a path that you can move quickly across without obstructions. Another important consideration is not to have escape routes headed up steep slopes. Remember, fire travels uphill fast. Trying to outrun a fire uphill with your fire gear is almost impossible.

The national standard is to identify escape routes, through flagging, is to use pink flagging. Do not assume everybody is aware of the location of the escape routes. Make sure they are flagged and walk them out to make sure they do not have obstructions. If they do, then try to move them or pick a new route.

A vehicle is a very effective tool to use for escape. Always make sure your vehicle is pointed toward the escape route. A safety zone may be further away and considered adequate, if all firefighters are mobile via their vehicles.

Minimum of Two Escape Routes at all Times!

© Grisha Bruev/Shutterstock.com

This illustration above would not be considered an adequate or appropriate escape route because of the amount of debris obstructing a potential retreat.

Courtesy of Brian Henington

This dozer line would be considered an adequate escape route to a safety zone.

[10] Wildland Fire Lessons Learned Center, *Six Minutes for Safety. Escape Routes* (Tucson: Wildland Fire Lessons Learned Center, National Advanced Fire and Resource Institute (NAFRI)), accessed February, 22, 2015.

[11] Ibid., accessed February, 22, 2015.

Safety Zones

A **safety zone** is defined as a preplanned area of sufficient size and suitable location that is expected to protect fire personnel from hazards without using fire shelters. The National Wildfire Coordinating Group provides an explanation of a safety zone as:

> *An area cleared of flammable materials used for escape in the event the line is outflanked or in case a spot fire causes fuels outside the control line to render the line unsafe... Safety zones may also be constructed as integral parts of fuel breaks; they are greatly enlarged areas which can be used with relative safety by firefighters and their equipment in the event of blowup in the vicinity.*[12]

Safety zones should be big enough to avoid the use of a fire shelter. Natural areas like a large river, lake, meadow, or rock outcropping can make good safety zones (if they are big enough). Man-made areas that make good safety zones could be large highways (if traffic has been blocked), agriculture fields, golf courses, etc. Constructed safety zones during fire activities can include areas such as helispots, clearcuts, fuel breaks, etc. The preferred safety zone along the fireline is the black or area already burned by the fire. There are some considerations when using the black as a safety zone (discussed later).

Safety Zones should not be located:

- Downwind from fire
- In chimneys, saddles and/or narrow canyons
- On steep slopes
- Heavy fuels
- Steep uphill escape routes.

Examples of safety zones:

Courtesy of Brian Henington

This burned meadow above would be an ideal safety zone if it is determined to be big enough.

Courtesy of Brian Henington

An overgrazed or drought-stricken pasture shown above would also make an adequate safety zone.

[12] National Wildfire Coordinating Group, *Glossary*, 153.

Courtesy of Brian Henington

The green agricultural field would be an ideal safety zone.

Courtesy of Brian Henington

This area is considered "Good Black" as it lacks available fuels to reburn. Trees should be checked to ensure they are not additional hazards. If so, the area could be used as a safety zone.

The size of safety zones should be based on the separation distance between the firefighter and the flames. There is a general guideline in the *Incident Response Pocket Guide* on separation distance; however it should be used as a guideline only. Again, your job as a Firefighter Type 2 is not to decide the location of the safety zone or the size of the zone. You should know that you must have an adequate safety zone before you engage a fire. Has your boss constantly evaluated the safety zone as your crew progresses and are you aware of its location?

Courtesy of Brian Henington

Crews are using Good Black as a safety zone for themselves and their equipment/vehicles.

A **deployment site** or zone is not a safety zone. A deployment site is an area used by firefighters to deploy a fire shelter or an area identified as a suitable location to deploy a fire shelter. The deployment site is specifically tied to a fire shelter and should not be considered a safety zone because it lacks the sufficient size to protect firefighters and equipment.

GOOD BLACK

an area that has burned available fuel and lacks more fuel for consumption, including the overstory or brush (which is considered reburn potential).

BAD BLACK

an area the fire has not burned completely or has preheated the overstory in such a way that the trees or brush could sustain fire movement and combustion.

One Foot in the Black

The concept of One Foot in the Black or Bring the Black with You describes the philosophy of using the black as a primary safety zone. If grass fires are attacked from the black or burned area, then the safety zone is always in place; in fact we are attacking from the safest location we can be in.

Personal Experience: In 1996, I was assigned to a district that was experiencing a very busy and severe fire season. We had numerous hotshot crews and strikes team assigned to support our fire operations. In addition, we also had a crew of smokejumpers assigned to us. The jumpers were assigned as a Type 1 Hand Crew. Some of these jumpers were on the South Canyon Fire and either survived the blowup or lost some of their friends on this tragedy fire.

Over a couple of weeks, we became very close to the jumpers. On one particular day, we were assigned to a fire with the jumpers. The fire began to increase in activity and intensity. On this fire, I learned a valuable and lifelong lesson from one of the jumpers named Tony Petrilli.[13] As we were digging fire line, I overheard the jumpers constantly mentioning the term, "Good Black" and "Bad Black." Although I had several years of experience, I had never heard the term before. I asked Tony what the term meant, without trying to sound like a dumb butt. Tony explained "Good Black" as an area that has burned available fuel and lacks more fuel for consumption, including the overstory or brush (which is considered reburn potential). He also explained "Bad Black" as an area the fire has not burned completely or has preheated the overstory in such a way that the trees or brush could sustain fire movement and combustion.

Tony taught me to determine if the black is adequate or if it is dangerous. Does it have reburn potential? If it does, then it is not considered Good Black and should not be used as a safety zone.

COMMON DENOMINATORS OF FIREFIGHTER FATALITIES

Fire experts have extensively studied historical wildland fire fatalities and entrapments. They have studied, analyzed, and researched the factors that have contributed to fire fatalities. Their goal was/is to develop standards or protocols for future firefighters to use so they can avoid a similar outcome. The first such effort was conducted in the 1970s by fire scientist Carl Wilson. Mr. Wilson used historical data to compare fire behavior elements associated with firefighter deaths. The denominators he found associated with fire behavior and fatalities have created and established the Common Denominators of Firefighter Fatalities that are widely used today throughout the wildland fire service to manage risk.

The Common Denominators of Firefighter Fatalities are identified below.

#1 On relatively small fires or deceptively quiet areas of large fires.

Why is it dangerous?

- Lack of adequate or working communication.
- Complacency.
- Lack of focus.
- Sudden increase in fire behavior.
- Fire activity elevates to the third dimension.

[13] Anthony (Tony) Petrilli is a highly respected firefighter and former smokejumper. He is currently an equipment specialist for the fire and aviation and safety and health programs at the Missoula Technology and Development Program. He has been involved in key fire safety related issues and technology development over his career. He is a qualified Division/Group Supervisor and Type 3 Incident Commander.

How to avoid:

- Monitor weather and fire behavior at all times.
- Every fire, regardless of size or intensity, has the potential to become extreme.
- Stay focused! Ensure you and your crew members are alert—Look Up, Look Down, and Look Around at all times.
- Ensure communication measures are in place and working. If they are not, correct the issue or disengage until adequate measures are in place.

Courtesy of Brian Henington

#2 In relatively light fuels, such as grass, herbs, and light brush.

Why is it dangerous?

- Dry fuels are impacted by drying and daily temperature variations.
- Wind has major influence on fire behavior in light fuels.
- Firefighter attitude of "Oh, It's Just a Grass Fire![14]
- Grass fires have the potential to burn very intense and move rapidly.
- Fire behavior can increase in a matter of seconds or minutes.

© James Bostian

How to avoid:

- On grass fires, attack the fire from the black.
- Monitor weather and fire conditions.
- Treat every grass fire as if it has the potential to produce extensive flame lengths and produce high rates of spread.

#3 When there is an unexpected shift in wind direction or wind speed.

Why is it dangerous?

- Wind has the greatest impact on fire behavior.
- Inactive fire behavior can suddenly become very active.

© Peter Weber/Shutterstock.com

[14] Concept developed by the Wildland Fire Lessons Learned Center to address the concerns and dangers associated with grass fires.

How to avoid:

- Monitor wind speed and direction at all times.
- Ensure you have a weather forecast.
- Evaluate available fuel.
- DO NOT ATTACK THE FIRE AT THE HEAD or in front of it.
- Ensure you have adequate safety zones and make them known.
- In light fuels, attack from the black.

#4 When fire responds to topographic conditions and runs uphill.

Why is it dangerous?

- Fire spreads uphill faster than downhill.
- Fuels are closer to the advancing flames.
- Convective and radiant heat are driven upslope, preheating available fuels.
- If firefighters are above the fire, they will be hampered by smoke, making visual identification of fire behavior almost impossible.
- Fire activity will increase during the burn period as wind and temperature increases.

How to avoid:

- DO NOT BE ABOVE A FIRE WITH FIRE BELOW!
- Attack from the bottom of the slope or hill where you can escape quickly, be able to view fire behavior, and will remain out of the worst smoke.

THE STANDARD WILDLAND FIRE ORDERS

The Standard Fire Wildland Fire Orders, or Fire Orders, are our doctrine of safety measures that guide our firefighting efforts. These rules cannot and should not be compromised. They are general rules with a broad approach to safety. In addition, they are a functioning standard that must be adhered during suppression activities and re-evaluated throughout the day. These ten rules should be considered as valuable as LCES and should be used in conjunction with LCES to determine if our tactical actions are appropriate and deemed safe.

The Fire Orders are identified below. It is important that you understand, and not just memorize the rules. If I were on a SWAT Team and was told to learn and understand 10 rules that would keep me safe, then I would ensure that I do it. The same approach applies to these rules. Learn and understand their role in safe firefighting and how these rules are considered the Rules of Engagement for wildland firefighters.

1. Keep informed on fire weather conditions and forecasts.
   - Weather and wind have the greatest impact on fire behavior.
   - Receive morning weather briefing report. Ensure others have the same weather forecast as you.
   - Monitor weather using your belt weather kit or Kestrel throughout the day. Monitor trends. Is relative humidity decreasing? Is temperature increasing? Are winds picking up? All of these factors will contribute to increased fire behavior.
   - Look for weather indicators: Increased cloud cover, wind direction/intensity, dust devils, etc.
   - Do not assume somebody else is monitoring weather conditions. All firefighters must monitor and communicate weather changes and conditions.

2. Know what your fire is doing at all times.
   - Monitor the fire behavior: What is the color of the smoke? What are the flame lengths?
   - To know what your fire is doing, you must be able to see it.
   - Lookouts are essential to this rule for monitoring fire activity and informing their crews.
   - Pay attention to rate of spread of the fire (how fast is it moving?), direction of spread, available fuel, topography influences, spot fires, etc.

3. Base all actions on current and expected behavior of the fire.
   - The burn period will change fire behavior. Anticipate the change. Plan accordingly.
   - Is LCES in place?
   - Consider the Fire Behavior Triangle: weather, topography, and fuels. What will these influences have on fire behavior?
   - Do not get tunnel vision. Fire at 1000 hours will act differently than the same fire at 1500 hours. Is your boss considering this change? Even if it does not occur, remember, it can, so plan accordingly.
   - Take a tactical pause prior to the burn period to evaluate current tactics.

4. Identify escape routes and safety zones and make them known.
   - Are safety zones adequate? Are they flagged? Has your boss shown you the location(s) of the safety zone(s)?
   - Do not assume somebody else will flag escape routes. Flag them!
   - Walk your escape routes out. Time your escape. Are your escape routes and escape time appropriate?
   - Make sure you have removed obstructions in the escape routes.
   - Are the escape routes and safety zones you are visiting this morning still going to be appropriate during the burn period?

5. Post lookouts when there is possible danger.
   - POST LOOKOUTS AT ALL TIMES. Every fire is potentially dangerous.
   - Is the lookout experienced? Is he or she good communicators?
   - Do you know the location of your lookout? Who is he/she?

6. Be alert. Keep calm. Think clearly. Act decisively.
   - Maintaining control of our emotions is one of the hardest things we can do in firefighting. The job is stressful. We are overloaded with information at times. Not getting information can also cause stress.
   - If you feel overwhelmed, there is absolutely nothing wrong with stopping and counting to 10.
   - Take a tactical pause if you are unsure of what you are supposed to be doing.
   - Experience and training will help with your stress levels.
   - Find a good and trustworthy mentor to help you learn to deal with stress.
   - The bottom line is that the decisions we make on a fire are often what leads to injuries or even worse. Learning to control our emotions is a skill that we all need to work on.

7. Maintain prompt communications with your forces, your supervisor, and adjoining forces.
   - Immediately respond to somebody calling you on the radio.
   - If you are busy, respond to them by stating "Stand by." This allows the caller to know that you are busy at the moment but will call back as soon as possible.
   - If you do not respond, then the caller is assuming something has happened to you.
   - Have extra batteries and make sure you get in the habit of establishing communications by conducting radio checks.
   - If adequate communications are not in place, then disengage the fire until they are in place.

8. Give clear instructions and be sure they are understood.
   - If we are responsible for our own safety, then we need to ensure we understand the instructions.
   - If you do not understand the instructions given to you, then ask for clarification.
   - If you supervise people, make sure you give adequate instructions. Allow firefighters time to ask for clarification.

9. Maintain control of your forces at all times.
   - You can ensure this rule is in place by knowing what your fire is doing at all times.
   - Ensure communications is established and functioning.
   - Is LCES in place and functioning?
   - If you feel like you are losing control of yourself or your employees, then take a tactical pause to regain control.
   - Regaining control does not mean screaming and cussing at people.

10. Fight fire aggressively, having provided for safety first! This is the most important rule of the Standard Firefighting Orders.
   - We have to be aggressive, but only when safety measures are addressed and implemented.
   - Ensure the fire is sized up and scouted prior to engaging resources.
   - Do you know what your assignment is? If not, get clarification.
   - Ensure LCES is in place.

The National Wildfire Coordinating Group summarizes the Standard Firefighting Orders the best with the following quote, "Before fully engaging, the Standard Firefighting Orders must be fully considered. If a safety problem arises at any point during engagement, then stop and re-evaluate the situation. Safety is written throughout the Standard Firefighting Orders. Always be prepared to re-evaluate and follow the sequence back through the orders at any time."[15]

Always remember why these rules were created. It is sad but reality; a firefighter or multiple firefighters have died because these rules were not established or followed.

THE 18 WATCH OUT SITUATIONS

The **Watch Out Situations** or the "18s" were created based on the evaluation of tragedy fires and the events that occurred that resulted in firefighter fatalities. These are guidelines and should be in place at all times. Unlike the Standard Firefighting Orders, the Watch Out Situations can be compromised, if and only if, mitigation measures are in place. The Watch Out Situations should be considered warnings. Alert somebody if you know you or your crew are in violation. Mitigation measures can be enacted quickly, or may require another tactical consideration that may take time to implement. "The response to the Watch Out Situations is to take positive action which will minimize the potential for serious injury or death."[16]

Sound the alarm if these guidelines are violated or are not in place.

The 18 Watch Out Situations

1. Fire not scouted and sized up.
2. In country not seen in daylight.
3. Safety zones and escape routes not identified.
4. Unfamiliar with weather and local factors influencing fire behavior.
5. Uninformed on strategy, tactics, and hazards.
6. Instructions and assignments not clear.
7. No communication link with crewmembers/supervisors.
8. Constructing line without safe anchor point.
9. Building fireline downhill with fire below.
10. Attempting frontal assault on fire.
11. Unburned fuel between you and the fire.
12. Cannot see main fire, not in contact with anyone who can.
13. On a hillside where rolling material can ignite fuel below.
14. Weather is getting hotter and drier.
15. Wind increases and/or changes direction.
16. Getting frequent spot fires across line.
17. Terrain and fuels make escape to safety zones difficult.
18. Taking a nap near the fire line.

[15] National Wildfire Coordinating Group, *Firefighter Training, S-130*, 4A.25.

[16] Ibid., 4A.6.

The Watch Out Situations are described below:

1. Fire not scouted and sized up.
   - » Hazards/Concerns:
     - Not able to identify hazards, values, fire potential, fuels, etc.
     - No plans in place before you engage the fire.
   - » Mitigation Actions:
     - Scout the fire. Walk the fire.
     - Conduct a valid size up.
     - Use the Size Up checklist in the IRPG.
2. In country not seen in daylight.
   - » Hazards/Concerns:
     - Have you ever gone camping when you arrived at your prime spot during darkness and you wake up in the morning to find another camper 200 yards from you? Darkness impacts your ability to see and base your actions accordingly.
     - Darkness impacts your ability to effectively size up a fire including identifying hazards.
   - » Mitigation Actions:
     - Can you wait to suppress the fire until daylight?
     - If not, proceed with caution.
     - Ensure LCES is identified and adequate.
3. Safety zones and escape routes not identified.
   - » Hazards/Concerns:
     - You do not know where the safety zones are located.
     - Not familiar with the escape routes.
   - » Mitigation Actions:
     - Make them known.
     - Walk out the escape routes and evaluate escape time.
     - Flag both safety zones and escape routes.
     - Brief other crews and resources on their location.
     - Do not engage until you have identified them.
4. Unfamiliar with weather and local factors influencing fire behavior.
   - » Concerns/Hazards:
     - Where I live, we have strong east winds that occur during cold fronts. The winds channel through a canyon and are released into adjancent areas with great velocity.
     - Certain areas will have specific weather factors that influence fire behavior.
   - » Mitigation Actions:
     - Ask a local firefighter. Get information on local weather patterns.
     - Once you receive an assignment in an unfamiliar area, start studying local weather events that impact the area.
     - Receive a weather forecast and monitor weather conditions at all times.

5. Uninformed on strategy, tactics, and hazards.
   » Hazards/Concerns:
      – Strategy is the overall plan to meet the objectives of the incident.
      – Some parts of the country or fire agencies will use different terminology. Again, attach yourself with a local firefighter to ensure you understand their terminology.
   » Mitigation Actions:
      – Ask for clarification.
      – Reference your IRPG if you are unfamiliar with the tactics or strategy.
      – Do not continue actions until you have rectified the issue(s) or feel comfortable implementing the tactics or strategy.

6. Instructions and assignments not clear.
   » Hazards/Concerns:
      – Do not know what you should be doing or how it should be done.
   » Mitigation Actions:
      – Ask for clarification.
      – Give clear instructions.
      – Do not engage the fire until your questions have been adequately addressed.
      – Give time for the crew to ask questions.

7. No communication link with crewmembers/supervisors.
   » Hazards/Concerns:
      – No communication with your crew or supervisor.
   » Mitigation Actions:
      – Establish communication and have a backup plan in place.
      – Stop activities if communication measures have been broken or are not in place.

8. Constructing line without safe anchor point.
   » Hazards/Concerns:
      – The fire has the potential to outflank you.
   » Mitigation Actions:
      – Always anchor from a safe and advantageous location.
      – Never attack a fire without safe anchor points.

9. Building fireline downhill with fire below.
   » Hazards/Concerns:
      – You cannot adequately see the fire.
      – Fire spreads uphill faster than downhill.
      – The convective energy is being released at the top of the hill.

» Mitigation Actions:
  - Walk to the bottom of the hill and anchor from the bottom and begin attacking the fire on the flanks. You can always see the fire if you use this concept.
  - There are established mitigation measures that can be used if downhill line construction is necessary. You can locate them in your IRPG—Downhill Checklist.
  - Implement LCES.

10. Attempting frontal assault on fire.
  » Hazards/Concerns:
    - No anchor point or adequate anchor point.
    - Chance of being outflanked by the fire is greatly increased.
    - You are at the head of the fire—the most dangerous part of the fire.
    - Smoke and heat will have a major impact on you.
  » Mitigation Actions:
    - Do not do it.
    - A time you may have to do this is the protection of structures; however, you will not do so if you are violating LCES.
    - If LCES is not in place, do not do it.

11. Unburned fuel between you and the fire.
  » Hazards/Concerns:
    - Allows the fire to hit you or your fireline with full force.
    - Fire direction can shift or change direction based on wind. The more fuel between you and the fire allows the fire to gain more energy.
  » Mitigation Actions:
    - Burn out the fuels following standard burnout guidelines and practices.
    - Direct attack can eliminate unburned fuels if direct attack can safely be done.
    - Implement LCES.

12. Cannot see main fire, not in contact with anyone who can.
  » Hazards/Concerns:
    - You cannot see the enemy.
    - Are not aware of fire behavior, rates of spread, etc.
    - Do not know when you should retreat or move to a safety zone.
    - Common in indirect attack.
  » Mitigation Actions:
    - Move to a location where you can see the fire.
    - Post one or multiple lookouts.
    - Proceed with caution.
    - Implement LCES.

13. On a hillside where rolling material can ignite fuel below.
    - Hazards/Concerns:
        - Rolling material can start new ignitions that may be below you, causing the fire to run uphill, which will jeopardize your escape routes.
    - Mitigation Actions:
        - Construct a cup trench (explained in later chapters).
        - Is there another location from which you can attack the fire?
        - Implement LCES.
14. Weather is getting hotter and drier.
    - Hazards/Concerns:
        - Increase in fire behavior.
        - Burn period.
        - Decrease in relative humidity and fuel moisture.
    - Mitigation Actions:
        - Implement and reevaluate LCES.
        - Observe weather.
        - Take weather readings on the hour. Monitor trends.
        - Observe fire behavior and communicate what you are seeing.
15. Wind increases and/or changes direction.
    - Hazards/Concerns:
        - Change fire spread and increase fire behavior.
        - More oxygen available for the fire.
        - Spotting chances are increased.
        - May increase reburn potential.
    - Mitigation Actions:
        - Monitor weather conditions.
        - Monitor trends.
        - Is LCES in place and has it be reevaluated?
        - Take a tactical pause. Reevaluate your tactics.
        - Ensure all personnel are aware of increased wind activity.
16. Getting frequent spot fires across line.
    - Hazards/Concerns:
        - Spot fires are an indicator of extreme fire behavior.
        - Fire brands below you can cause new ignitions which will spread uphill.
        - Spot fires may hamper suppression activities and increase danger to firefighters.
        - Spot fires can burn back to the main fire which creates major concerns if firefighters are between the two.

» Mitigation Actions:
  - Routine patrols to identify spot fires.
  - Immediately deal with them.
  - Numerous spot fires are an indication that extreme fire behavior is occurring.
  - Is LCES in place and has it been reevaluated?
  - Take a tactical pause and let fire managers know of the frequent spot fires.
  - Retreat to a safety zone.

17. Terrain and fuels make escape to safety zones difficult.

» Hazards/Concerns:
  - You do not want obstructions in your path. You want to be able to escape quickly, not hurdling trees, logs or rocks.
  - May increase escape time.
  - Increased potential of firefighter injury trying to navigate through dangerous terrain.

» Mitigation Actions:
  - Walk escape routes prior to needing them. Are they clear of obstructions?
  - Flag escape routes. At night, use glow sticks.

18. Taking a nap near the fire line.

» Hazards/Concerns:
  - What? I thought we are getting paid to work? There are times that fire managers allow us to rest. If you do so, make sure you are not on the fireline.
  - A majority of activities occur on the fireline: heavy equipment, engines traveling, etc.
  - The risk of falling trees, snags, and widow makers is increased along the fireline.

» Mitigation Actions:
  - Ensure your boss knows your rest location and you have received permission to rest. Do not rest alone; make sure you are with at least one other firefighter.
  - Locate a safe location to rest.

The Watch Out Situations should be used at all times by firefighters. If the guidelines are compromised, then provide mitigation. If mitigation measures are not adequate or applied, then do not continue your firefighting efforts until they are. This concept may require that you disengage the fire and recalculate the appropriate tactical methods to suppress the fire.

SUMMARY

Wildland firefighting is dangerous. It is complex, dynamic, and ever changing. Over 1,000 firefighters have died while they were attempting to suppress a wildland fire. Many of these tragedies could have been avoided if the Rules of Engagement were in place and prioritized. This chapter discusses 36 rules or guidelines that will keep you safe on a wildland fire. The rules can be difficult for entry-level firefighters to understand. If you use the concept of LCES, you will utilize a proven and effective safety foundation. Once you understand LCES, then you can begin to apply the other 32 rules or safety guidelines. Concentrate on the four LCES concepts (Lookouts, Communications, Escape Routes, and Safety Zones). Once you do, then the Standard Fire Orders will make sense and you will begin to understand why and how the rules are applied. You should begin studying, learning and comprehending the orders. Once you understand those rules and can recite them, then use the Common Denominators of Firefighter Fatalities as a tool to ensure you or your crew is not in violation of any of the denominators. Finally, concentrate on why the Watch Out Situations are dangerous and why not using them will compromise your safety. Know the mitigation steps and concentrate of studying them with your fellow firefighters.

The foundation of safety is our individual responsibility. Do not rely on your supervisor to keep you safe. Learn the Rules of Engagement and apply them to every fire you ever fight. If you use these proven concepts, you will return safely at the end of each shift.

KNOWLEDGE ASSESSMENT

1. Define a safety zone.
2. Who should function as a lookout?
3. What is the preferred safety zone for firefighters? What would be the determining factor if the safety zone would be considered adequate?
4. Identify the elements involved in LCES.
5. Who did Paul Gleason suggest should be a lookout on an Interagency Hotshot Crew?
6. What is considered Good Black? What is considered Bad Black?
7. Define a deployment site. Why is it different from a safety zone?

ACTIVITIES

1. Using your *Incident Response Pocket Guide (IRPG)*, locate LCES and specifically lookouts. What does the IRPG recommend for a lookout?
2. Locate the Standard Firefighting Orders in your IRPG. On what page can you locate these rules? You should begin to learn and understand the rules immediately. Begin quizzing yourself. Truly understand their intent. You should approach this as if you are going to teach a new group of firefighters how important these rules are.
3. Research the South Canyon Fire Investigation Report. You can locate this document online by conducting a web search of South Canyon Fire Investigation Report. Locate the Witness Statement completed by Tony Petrilli. Mr. Petrilli lists the violations of the Watch Out Situations that occurred on this fire. How many were in violation? What do you think about the violations?
4. View the video produced by the Wildland Fire Lessons Learned Center "Oh, It's Just a Grass Fire." https://www.youtube.com/watch?v=hl1gNlF0JkY. Provide your instructor with a short narrative of the video.

National Wildfire Coordinating Group. *Firefighter Training, S-130.* Boise: National Wildfire Coordinating Group, 2003.

National Wildfire Coordinating Group. *Glossary of Wildland Fire Terminology.* Boise: National Wildfire Coordinating Group, 2014.

Wildland Fire Leadership. *Leaders We Would Like to Meet—Paul Gleason.* http://www.fireleadership.gov/toolbox/leaders_meet/interviews/leaders_PaulGleason.html.

Wildland Fire Lessons Learned Center. *6 Minutes for Safety: Common Denominators of Fire Behavior on Tragedy Fires.* Tucson, Wildland Fire Lessons Learned Center, National Advanced Fire and Resource Institute (NAFRI), 2013. http://www.wildfirelessons.net/viewdocument/?DocumentKey=6b5bae01-b7c2-4ec2-918d-563005388def&tab=librarydocuments.

Wildland Fire Lessons Learned Center. *Six Minutes for Safety. Escape Routes.* Tucson: Wildland Fire Lessons Learned Center, National Advanced Fire and Resource Institute (NAFRI). http://www.wildfirelessons.net/viewdocument/?DocumentKey=a66917f9-afce-435d-938e-e53f09eeef72&tab=librarydocuments.

Wildland Fire Lessons Learned Center. *Video: Oh, It's Just a Grass Fire. Firefighter: Remember This Series.* Tucson: Wildland Fire Lessons Learned Center, National Advanced Fire and Resource Institute (NAFRI), 2011. https://www.youtube.com/watch?v=hl1gNlF0JkY.

CHAPTER

CHAPTER 9 LAST RESORT SURVIVAL

Learning Outcomes

- Explain the concept of last resort survival and how it applies to wildland firefighting.
- Understand how an escape fire can be an adequate last resort survival technique.
- Translate how structures or vehicles can be used a temporary refuge from an advancing fire.
- Compare the difference between a safety zone and a deployment site.
- Describe the two important protection features associated with the New Generation Fire Shelter.
- Explore deployment site features and explain how they are beneficial or detrimental to a firefighter.

Courtesy of Brian Henington

Key Terms

Burnover
Deployment Site
Entrapment
Entrapment Avoidance
Escape Fire
Fire Shelter
Last Resort Survival
New Generation Fire Shelter
Survivable Area
Trigger Point

OVERVIEW

Dynamic changes to fire behavior and intensity can create situations that eliminate escape options and can entrap firefighters. Firefighters' options to escape may be greatly reduced or eliminated. Firefighters must take immediate action(s) to protect themselves. This is what wildland fire managers refer to as Last Resort Survival. This concept may involve a fire shelter or may involve another escape method.

This chapter explores last resort survival options and introduces you to **fire shelters**. Fire shelters are part of required personal protective equipment and should be available at all times. If you are responsible for your own safety, then you need to familiarize yourself and practice the techniques identified in this chapter or in other reference materials. These techniques can save your life.

LAST RESORT SURVIVAL

Last Resort Survival can be explained as: All available and reasonable options for an escape to safety from a wildland fire have been exhausted. Firefighters must take other appropriate actions to survive. This typically involves the use of a fire shelter. The key to any successful fire operation is be sure that available escape options are not eliminated or compromised. We should be able to witness the progression of fire behavior to extreme burning and take appropriate actions before our options are eliminated. We should always use the Rules of Engagement to ensure our safety is not compromised. Once a violation or infraction of the Rules of Engagement occur, then disengage the fire and retreat to a safety zone. Take a tactical pause and determine what actions, if any, can be safely taken to suppress the fire. Remember, 100% disengagement from a fire is a reasonable and allowable tactical option.

This section introduces you to techniques that you can use to escape an advancing wildland fire. Although we focus on entrapment avoidance, you should understand that fire behavior can escalate very rapidly and your options for escape may be eliminated. You can also use your *Incident Response Pocket Guide* to familiarize yourself with some of these concepts covered here. Know what page the techniques are located on and practice these escape options during training exercises—without live fire. Constant training on these concepts will immediately spring into action if you have to use them.

Firefighters can use several proven techniques to avoid an entrapment or a burnover. A proven concept is the use of trigger points. **Trigger points** are "geographic points on the ground or specific points in time where an escalation or alternative of management actions is warranted."[1] An example of effective trigger points use is: If the fire reaches Point A (pre-established and known to all firefighters); then we must retreat to a safety zone. Trigger points are highly effective and should be used at all times by firefighters to avoid entrapments.

[1] National Wildfire Coordinating Group, *Glossary of Wildland Fire Terminology* (Boise, National Wildfire Coordinating Group, 2014), 118.

Entrapment A situation when personnel are unexpectedly caught in a fire behavior-related, life-threatening position where planned escape routes or safety zones are absent, inadequate, or compromised. An entrapment may or may not include deployment of a fire shelter for its intended purpose. These situations may or may not result in injury. (NWCG. 2014).

Entrapment Avoidance A process used to improve the safety of personnel on the fireline, which emphasizes tools and tactics available to prevent being trapped in a burnover situation. This process includes appropriate decision making through risk management, application of LCES, use of pre-established trigger points, and recognition of suitable escape routes and safety zones. (NWCG. 2014).

The concepts below can be used when escaping a fire with or without the use of your fire shelter.

1. Try all options to escape the advancing fire. This may include outrunning the fire, using vehicles, helicopters, or taking refuge in a lake or large river.
2. What is your best available option? Use it! Seeking refuge in a lake may be appropriate if the lake is deeper than two feet. Make sure firefighters can swim.
3. When outrunning or moving away from the fire, drop your gear. Your fire gear can slow your retreat time considerably. If you do run, you must keep your fire shelter, water (pull out your camelback or grab a canteen), radio, gloves and hand tool.
4. Look for light fuels, such as grass. Light fuels will not produce as much heat as heavy fuels. The residual burning of light fuels is far less than heavy fuels.
5. A fire shelter could be used as a heat barrier while you are moving. Extensive practice with this technique is highly recommended.

If you are not able to escape the fire by using any of the steps above, then you need to consider locating a **survivable area**. It is important to note, the survivable area may require the deployment of a fire shelter to survive. Below is a list of recommendations for a survivable area:

- Stay away from chimneys, chutes, saddles, or other hazardous terrain features.
- Locate lakes, ponds, or rivers. They should be more than two feet deep.
- Pull a "Wag Dodge." (See below.) Ignite an escape fire in light fuels.
- In heavy fuels, ignite a backfire.
- Call for air support: helicopter bucket drops or airtanker retardant drops.
- If there is time, cut and scatter available fuels.
- Look for heat barriers: Large rocks, dozer berms, etc.
- If you use a road, consider if there is fire traffic that could run over you. If so, roads would not be recommended. Large roads may be suitable if the shoulders of the road are large enough.
- Seek temporary refuge in a vehicle or structures (explained later).

Examples of areas that may be suitable for survival

© Mykola Mazuryk/Shutterstock.com

This river may be a suitable survivable area.

© Topseller/Shutterstock.com

Is this lake suitable to survive a fire? Can you or your other crew members swim?

Courtesy of Brian Henington

Helicopters may be able to extract you before the advancing flames hit your location. Don't count on them at all times. Have other options in place in case the helicopter cannot land at your location.

© Bogdanhoda/Shutterstock.com

Dozer berms may provide relief but you may have to deploy a fire shelter to survive. Follow your supervisor's orders. Always deploy a shelter if you feel like you need to.

Escape Fire in Light Fuels: The Wag Dodge

In 1949, several smokejumpers were assigned to a wildland fire in northwestern Montana. The fire, known as the Mann Gulch Fire, was burning in steep terrain with the dominant fuel consisting of continuous, very dry grass intermixed with trees. As the firefighters were working on the fire, a sudden increase in fire behavior caused the fire to elevate to extreme burning conditions. The firefighters used an uphill escape route to try to outrun the fire. As the firefighters were progressing uphill, the smokejumper foreman, Wag Dodge, stopped running and lit the grass in front of him on fire. He did this before the main advancing fire had reached his location. Wag Dodge effectively lite an escape fire. As his escape fire burned uphill, (knowing that fire will travel uphill), Wag followed the flames as a running back in the NFL would follow his offensive line or blockers.

When the main fire reached the point of ignition, Wag lay on the ground (farthest away from the advancing fire) and was able to survive the fire. Wag and two other firefighters survived. Twelve other firefighters died. Not one firefighter used Wag's escape fire except himself. The other two survivors were able to survive the fire on a rock slide. Wag's quick and effective decision was the first documented event of a firefighter lighting an escape fire during active fire suppression activities.

When you explore this concept, think about how quick and decisive Wag's actions were. The other survivors thought Wag was crazy for starting a new fire when a fire was racing uphill at them with exceptional speed. Did Wag understand that they were not going to outrun a fire uphill? Did he also understand that grass fires are going to move very fast, but after they have burned there is limited residual burning? His knowledge and understanding of fire behavior allowed him to use fire as an asset for his safety, not a detriment to his survival.

Vehicles as Escape Mechanisms

Vehicles may be the ideal tool to use when escaping an advancing fire. If you are ever caught in underburned fuel or in light fuels with an advancing fire approaching you, turn your vehicle into the advancing flames and drive into the black. Once you pass the initial heat blast, the fuels on the other side of the flames will be consumed. Make sure your windows are up and the air conditioner is turned off.

Below are key concepts when using your vehicle as an escape:

- Always point your vehicle toward your escape route. You do not want to lose valuable time backing up a vehicle when you need to escape.
- Close all doors, windows, close vents, and turn off your air conditioner.
- Engage your overhead emergency lights or hazard lights.
- Escaping firefighters may be using the road as an escape route and/or as a deployment site. Drive with caution.

If you have to use your vehicle as temporary refuge (you cannot drive away) do the following: (These measures should be done only in light fuels, not heavy fuels.)

- Close all the vents, turn off motor, and roll up the windows.
- Deploy your fire shelter and use as a radiant heat reflector on the windows.
- Lie down on the floor and face away from the advancing flames.
- Protect your airway.
- Once the initial flame front passes, get out of the vehicle and move away. Deploy your shelter in a safe location.[2]

---

[2] Information taken from the *Incident Response Pocket Guide*, 2014.

Author Note: My engine and two other firefighters survived a burnover by using the technique of driving through the flames into the black. In 1996, we were flanking a fast-moving grass fire north of Las Vegas, New Mexico. Our engine was involved in direct attack by attacking the fire from the black. As we progressed, we had to leave the black to move our engine around a large boulder (volcanic rock) field. We traveled close to 200 yards from the fire edge to find a suitable location to navigate our engine through the rocks. As we reengaged the fire, a thunderstorm several miles away caused a shift in wind direction. The flank of the fire now became the head of the fire. We could not retreat from the fire because of the large boulders and had to take quick and decisive actions. An escape fire was not an option as the flames were approaching too fast. At that point, some instructor's voice popped into my mind: "If you ever get caught in a fast-moving grass fire, drive your engine right through the flames into the black." We did so and were all safe (except one of the decals on the engine melted). Lesson Learned: Always attack a grass fire from the black.

Structure or Homes as Temporary Refuge

Homes or structures can provide a temporary relief from an advancing fire. Although the home may eventually burn down, it may provide relief from the initial heat blast produced by the fire. If a structure is identified as a possible escape option, you must consider some of the following:

- What is the building construction material? A brick or stucco house will not become involved in fire as a home with wood siding or wood shingles.
- The structure will present new and challenging hazards: Is the electricity on? Is the gas/propane still on? What hazardous materials are in or next to the house?
- Has the home been prepped and protected through firefighter actions?
- Trailer homes will burn the fastest. It is not recommended using trailer homes as a temporary relief from an advancing fire.
- If the home presents more hazards, then the structure may not be an ideal location to seek temporary refuge.

If the structure does not appear to have additional hazards, then consider using it to escape the initial heat blast. Below are some tactical considerations when using a structure as temporary relief.

- Burn out the fuels around the home if you have time.
- Remove flammable materials next to the home. Throw lawn furniture to the side away from the advancing flames.
- Enter the home:
  - » Close the windows and shut the blinds or shades.
  - » Stay away from windows.
  - » Go to the exit away from the advancing flames.
  - » Lie down on your belly, with your face down.
  - » Stay calm. After the initial blast and fire front passes, go out the exit the flames hit first.[3]

---

[3] Information taken from the *Incident Response Pocket Guide*, 2014

- Do not deploy a fire shelter in the structure.
  - » When you leave the home, stay low, move away from the structure and deploy your fire shelter far enough from the structure.

There has been a lot of discussion and debate on using structures as a temporarily refuge from a fire. The key term here is TEMPORARY. Remember, this may be your only option. The structure should not become fully involved in a matter of seconds. When you take refuge inside the structure, you may receive relief from the initial heat blast. This may save your life. Do not stay in the house while it is burning. Remember, a bandana is not going to protect your airway from toxic burning materials or chemicals.

If all other escape options have been eliminated, then we must prepare for a fire shelter deployment. Always remember: The most appropriate and immediate action to survive a fire may require that you deploy a fire shelter instead of trying the other escape options identified above. As they say, Better Safe than Sorry!

THE FIRE SHELTER

The fire shelter is mandatory personal protective equipment and should be worn by wildland firefighters at all times. The fire shelter should be easily accessible and carried in an approved fire shelter pouch, on a chest harness, or on your belt. The fire shelter is essential to survive a last resort situation. Fire shelters have saved hundreds of lives and should be considered an important part of our firefighting protective measures. "The fire shelter has been required equipment for wildland firefighters since 1977. Since that time, shelters have saved the lives of more than 300 firefighters and have prevented hundreds of serious injuries."[4]

Burnover An event in which a fire moves through a location or overtakes personnel or equipment where there is no opportunity to utilize escape routes and safety zones, often resulting in personal injury or equipment damage.

Images provided by the UD Forest Service: Missoula Technology and Development Center and the National Interogency Fire Centers

[4] Leslie Anderson, *The New Generation Fire Shelter* (Boise: National Wildfire Coordinating Group, 2003) 1.

New Generation Fire Shelters

The only fire shelter that is acceptable on federal and most state fires is the New Generation Fire Shelter. This fire shelter was created in 2003 and has some improved features to add additional protection for the firefighter. The National Interagency Fire Center provides the following explanation on the New Generation Fire Shelter.

> *The new generation fire shelter offers improved protection from radiant and convective heat. All federal, state, and local wildland firefighters carry the fire shelter while working federal fires. As always, the fire shelter is the last line of defense when facing a fire entrapment; escape is always the highest priority. Fire shelters will not guarantee a firefighter's survival in an entrapment situation. Firefighters should do everything they can to avoid situations where they need to deploy a fire shelter.*

Fire shelters are not intended for firefighters to take greater risks or to violate the Rules of Engagement. In our career, we should strive to never use a fire shelter because we have prioritized entrapment avoidance. According to Tony Petrilli, Fire Shelter Project Leader with the Missoula Technology & Development Center "Remember, do everything you can to avoid situations where you might have to use a fire shelter."[5] With that said, we also have to understand that in our world of dangerous and dynamic circumstances, you may have to use a fire shelter to save your life.

Figure 9-1 is taken from the National Interagency Fire Center and is intended to demonstrate how the fire shelter reflects radiant heat.

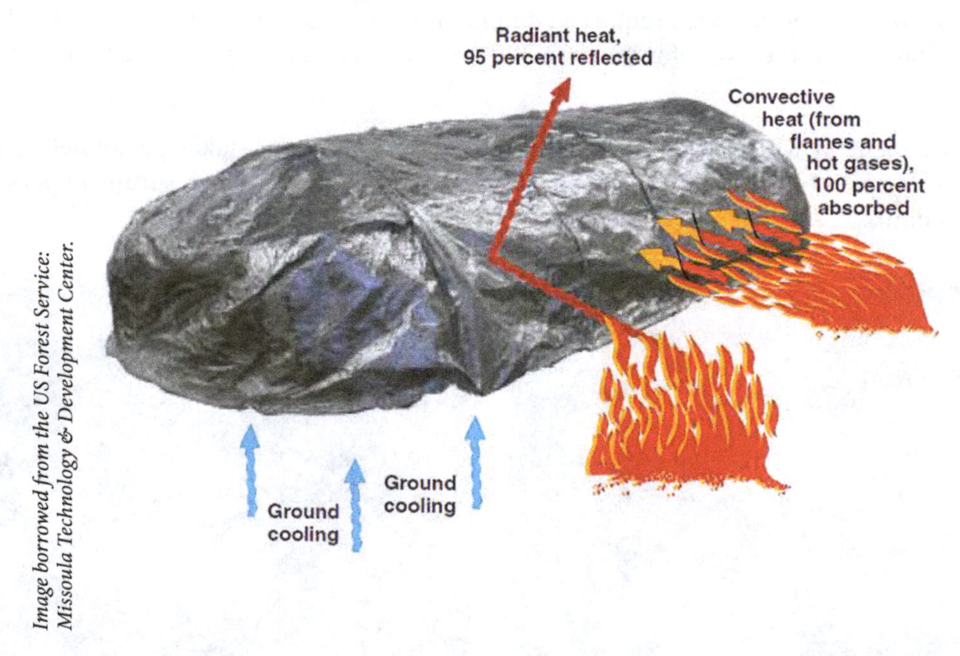

Image borrowed from the US Forest Service: Missoula Technology & Development Center.

Figure 9-1 *Reflection of Radiant Heat by the New Generation Fire Shelter.*

---

[5] United States Forest Service: Missoula Technology and Development Center, *Video: Introduction to Fire Shelter Deployment Stories*, accessed March 14, 2014.
https://www.youtube.com/watch?v=cFb-LVfZ8K8&list=PLTjug05B4KNt-OVXS8Ce93vqSufg-sOmY&index=1.2014.

The New Generation Fire Shelter is designed to provide two important protection features: 1) reflect radiant heat and, 2) trap breathable air for the firefighter. The outer layer of the fire shelter is made from aluminum foil that is bonded to woven silica. This material is designed to slow radiant heat movement from outside the shelter to the inside. The inner layer of fire shelter is made of aluminum foil laminated to fiberglass. This design is intended to "...prevent heat from reradiating to the person inside the shelter."[6] The two layers are sewn together which creates an air gap that provides extra insulation for the firefighter. The fire shelter is effective up to 500° F. If temperatures exceed 500⁰ F, than the fire shelter can begin to break down. Remember the shelter is not designed for direct flame contact.

HOW DOES THE NEW GENERATION FIRE SHELTER WORK?

1. Reflect radiant heat, and
2. Provide cooler, breathable air during the time of entrapment.

Your airway and lungs are the most important part of your body to protect when in a fire shelter. Many firefighters have survived third degree burns on their body; however surviving severe damage to your lungs and airways is almost impossible.

Fire Shelter Maintenance

It is our responsibility to maintain our fire shelter. We should ensure the fire shelter is functioning and not damaged. You can inspect your fire shelter by looking at the protective plastic case.

Some helpful hints for inspecting and caring for your shelter:

1. The protective plastic should be clear.
2. If the protective plastic is dark gray or black, you need a replacement shelter. The shelter material is breaking down.
3. It is recommended to inspect the fire shelter every two weeks for any obvious damage.
4. As a reminder, you never want to pull the red tabs to expose the shelter. If this is done, than the shelter is no longer available for use on a fire.
5. Fire shelters are very expensive; we need ensure we do everything to protect them.

Courtesy of Brian Henington

The fire shelter shown above should be replaced. Notice the dark grey and black spots.

[6] Anderson, *Fire Shelter*, 2.

Care and Handling

You should handle your fire shelter with care. The most common damage derives from firefighters sitting on their fire shelter because it is located on the bottom of their fire pack. You should be aware that can lead to damage to the fire shelter. In addition, the fire shelter should always be carried inside its enclosing the plastic liner. Finally, never carry a shelter inside the main pouch of your fire pack. Doing so would make the fire shelter useless.

Another helpful hint is to always carry an extra set of gloves next to your fire shelter. Most fire packs will have enough room in the fire shelter pouch to store a pair of gloves. Your hands can become exposed during a fire shelter deployment; therefore, gloves are a necessity. It is common for firefighters to lose their gloves during fire activities.

Deployment Sites

As a reminder, a fire shelter deployment site is not a safety zone. A safety zone should be big enough that you can survive a fire without the use of a fire shelter. A **deployment site** involves the use of a fire shelter. The National Wildfire Coordinating Group explains a deployment site as: "Deployment sites are used when fire conditions are such that escape routes and safety zones have been compromised. Deployment sites are last resort locations with generally light fuels in which a firefighter must deploy a fire shelter to aid in survival."[7] To escape to a deployment site, drop your gear and run. Keep your fire shelter, hard hat, radio, gloves, and drinking water. Fire packs will contain fusees that can burn at excessive temperatures and will cause additional harm to you.

Good fire shelter deployment sites may include:

- Low areas (lowest intensity and smoke) to deploy.
- Choose an area where the fire shelter has a good seal to the ground. Large rocks may cause an incomplete seal with the ground, which will allow smoke and superheated gases to enter your shelter.
- Look for heat barriers.
- Look for light fuels.
- Locate a rock outcropping with smaller rocks.

Bad fire shelter deployment sites may include:

- Chimneys.
- Chutes.
- Saddles.
- Steep Slopes.
- Near snags or widow makers.
- Near heavy fuels.
- Boulder fields with air gaps between the rocks and the ground.

Once you are inside the fire shelter: Protect Your Airways at All Costs!

[7] National Wildfire Coordinating Group. *Firefighter Training, S-130* (Boise: National Wildfire Coordinating Group, 2003), 4B.14.

Once you have selected a site, you should consider the following:

1. If you have enough time, clear a spot big enough for your fire shelter from any flammable fuels. Dig a hole for your head. (This will provide cooler air). Again, you will only do this if you have the appropriate time.
2. If you can, burn out an area of light fuels to deploy in.
3. If you can locate a large heat barrier, this may help reduce the blast of superheated air and gases as the fire reaches you.
4. If you choose a river, pond, or lake ensure the water is at least two feet deep. Deploy your shelter and place your head inside the shelter to create an airway.
5. Find a low spot and deploy inside of it.
6. If you do not have time to accomplish any of the above, GET INSIDE YOUR FIRE SHELTER IMMEDIATELY!

Getting Inside the Fire Shelter

1. ALWAYS POSITION YOUR FEET TOWARD THE ADVANCING FLAMES.
2. Ensure you are holding onto the shelter handles as tight as you can.
3. Get into your shelter before the advancing flame front hits your location.
4. Your face should be toward the ground. Lie prone.
5. Protect your airway at all times. Use a bandana or neck shroud.

Inside the Fire Shelter

1. Drink water, but make sure you don't pour the water on your clothes. This will create steam that damages your lungs if inhaled.
2. You should expect extreme heat.
3. Call for air support on the radio.
4. You should expect heavy ember showers and powerful winds. Hang onto your shelter.
5. Use your fellow firefighters as a source of encouragement and bravery. Encourage them to stay inside their fire shelters until it is safe to come out.
6. Survivors have said that the noise and the heat were unbearable; however they waited out the fire front and were able to survive.
7. Do not get out of your fire shelter too soon. A good rule of thumb is to wait until the rescuers tell you to come out. If rescuers are not coming, ensure temperatures and the heat are low enough that you can escape safely.

Fire Shelter Training Steps

The steps identified below should be used during training and during an actual fire shelter deployment. Training is essential for effective use of a fire shelter. Most agencies utilize practice fire shelters for training due to the cost of the actual fire shelter. The training shelter will provide the same function and steps of the real fire shelter in training exercises.

Note: The steps below are from a standing position only. Firefighters should practice deploying a shelter from their knees or on the ground in addition to this technique. The technique below is the easiest technique to teach entry-level firefighters. The step-by-step process uses a practice fire shelter. Practice shelters are for practice only. They should never be worn during fire suppression activities.

Courtesy of Brian Henington

Step 1: Gloves: If you have lost your gloves, put on your extra gloves stored in the fire shelter pouch next to your fire shelter.

Courtesy of Brian Henington

Step 2: PPE Check:

- Are your gloves on?
- Is your hard hat on with the chinstrap fastened around your chin?
- Do you have water and your radio?
- Put on your neck shroud and/or bandana.
- DO NOT WET YOUR CLOTHES. THIS INCLUDES YOUR BANDANA.

Step 3: Remove the fire shelter from the case. Throw away your fire pack and any flammables (if you still have your pack).

Courtesy of Brian Henington

Courtesy of Brian Henington

Step 4: Pull the red ring tabs from the plastic to expose the fire shelter.

Courtesy of Brian Henington

Step 5: Grab the yellow handles attached to the shelter. Grab the right handle (identified in red letters) with your right hand and the left handle (identified in black letters) with your left hand.

Courtesy of Brian Henington

Step 6: Hold tight and firmly. Shake the fire shelter. This technique allows you to capture essential air for breathing.

Courtesy of Brian Henington

Courtesy of Brian Henington

Step 7: Point your feet to the oncoming flames.

Courtesy of Brian Henington

Step 8: Put your right foot into the shelter. Right foot should be standing on your shelter—inside the shelter.

Courtesy of Brian Henington

Step 9: Make a hard uppercut with your right hand. This will expand or extend the fire shelter. At this point, your right hand should be in the fire shelter.

Courtesy of Brian Henington

Step 10: Pull the shelter over your head. This will also allow for your left arm and shoulder to enter the inside of the shelter.

Courtesy of Brian Henington

Step 11: Step into the shelter with your left foot.

Courtesy of Brian Henington

Step 12: Fall to your knees (with your feet toward the advancing fire).

Courtesy of Brian Henington

Step 13: From your knees, fall to your stomach.

Courtesy of Brian Henington

Step 14: Make sure your hands and feet are inside the straps built into the fire shelter.

Courtesy of Brian Henington

Step 15: Push out the fire shelter with your hands. Allow for air insulation. You do not want the shelter directly touching your body.

Step 16: If you have to move, do so by scooting on your stomach. Do not stand back up or crawl on your knees.

Step 17: Wait until rescuers tell you to get out or when you are convinced the fire threat is over. Do not come out of the shelter too early.

Step 18: All of the above steps should be completed in less than 25 seconds.

Continued Training

Once you have mastered the above technique, you should try different types of deployments.

- Practice deploying fire shelters on very windy days or by using a large fan to mimic the wind conditions involved in an entrapment.
- Try deploying from the ground while you are lying down.
- Deploy as groups. Practice using multiple firefighters deploying their fire shelters. This will help with deploying close together and improve accountability measures within a crew.
- Never attempt to use a practice shelter in live fire situation or carry on the fireline.
- Do not practice with live fire activities.
- Learn to use your shelter under the worst-case scenario and under stressful situations.
- Try running 200 yards, then deploy as a group.
- Become familiar with Leslie Anderson's training manual on the New Generation Fire Shelter: *The New Generation Fire Shelter.*
- Practice, Practice, Practice.

Courtesy of Brian Henington

Every firefighter should constantly practice fire shelter deployments; not just once a year during the mandatory annual fire refresher course. You should practice as if your life depended on it. If you were a SWAT team member, wouldn't you want to practice firing your weapon as much as possible? We should incorporate the same philosophy for practicing fire shelter deployments. Practice all the time, not just once a year. Practice makes perfect!

SUMMARY

Our job as effective firefighters is to avoid an entrapment situation. If we use the Rules of Engagement then we should have enough time to escape a burnover situation. Although this concept is the ideal approach to safety, we must understand that fire behavior can change very fast and can cut off your planned escape.

There is no firefighter in the world who is invincible! You should consider the techniques covered in this chapter as a tool for your firefighting toolbox to help you survive an entrapment or burnover. If you start quizzing yourself during training activities, then you will have a grasp on what a survivable deployment site is and one that is considered not survivable. Finally, you should practice fire shelter deployments as much as possible and become familiar with the different deployment procedures. Those who have survived using a fire shelter will stress to you how events can change drastically and how you may have to use a fire shelter to survive. They will also tell you how essential training is when deploying a shelter as a last resort survival option.

KNOWLEDGE ASSESSMENT

1. The key to any successful fire operation is not to have ______________________.
2. We should always use the ______________________ to ensure our safety is not compromised.
3. What are trigger points?
4. Who was the first documented firefighter to use an escape fire to survive an advancing fire? What was the name of the fire and what year did it occur? Where did it occur?
5. ______________________ may be the ideal tool to use when escaping an advancing fire.
6. What direction should your vehicle always be pointed?
7. Can you use a structure as a temporary refuge to escape an advancing fire? What advantage does this have?
8. Is the fire shelter considered mandatory PPE?
9. True or False: Fire shelters are not intended for firefighters to take greater risks or to violate the Rules of Engagement.
10. What two important features does the New Generation Fire Shelter provide?
11. What is the maximum temperature (in Fahrenheit) a fire shelter is considered effective? Is it designed for direct flame contact?
12. How often should you practice deploying your fire shelter?

EXERCISES

1. What concepts can be used when escaping a fire with or without the use of your fire shelter?
2. If you are ever caught in underburned fuel (in light fuels) with an advancing fire, turn your vehicle into the advancing flames and drive into the black. Explain why this concept is proven.
3. Are a safety zone and a deployment site the same thing? What is the difference?

BIBLIOGRAPHY

Anderson, Leslie. National Fire Coordination Center. *The New Generation Fire Shelter*. Boise: National Wildfire Coordinating Group, 2003.

National Wildfire Coordinating Group. *Firefighter Training, S-130*. Boise: National Wildfire Coordinating Group, 2003.

National Wildfire Coordinating Group. *Incident Response Pocket Guide*. Boise: National Wildfire Coordinating Group, 2014.

National Wildfire Coordinating Group. *Glossary of Wildland Fire Terminology*. National Wildfire Coordinating Group. Boise, Idaho. 2014.

United States Forest Service: Missoula Technology and Development Center. *Video: Introduction to Fire Shelter Deployment Stories*. Directed and Produced by the United States Forest Service: Missoula Technology and Development Center. Sponsored/Hosted by Wildland Fire Lessons Learned Center.

CHAPTER 10

FIRELINE HAZARDS

Learning Outcomes

- Define safety and explain a safety culture.
- Analyze firefighter fatalities in relation to specific fireline hazard types.
- Identify the two categories of fireline hazards.
- Explain the dangers associated with hazardous trees.
- Describe the process involved with refusing risk.

Courtesy of Brian Henington

Key Terms

Fireline Hazards
Hazard Trees
Mitigation
Objective Hazards
Safety
Snags
Subjective Hazard
Widow Makers

OVERVIEW

Fireline hazards are part of our working environment. They may range from environmental, biological, terrain related, and/or human activities. All of these hazards have the ability to cause injury or death to firefighters. It is our responsibility to ensure we are able to identify hazards, evaluate the exposure and threat to firefighters, and make the appropriate mitigating decision to avoid or eliminate the risk.

This chapter will focus on several different fireline hazards. It is important to note that not all hazards will be covered in this chapter. Some hazards may be specific to your jurisdictional area. You should become familiar with those hazards and identify mitigation measures.

FIRELINE HAZARDS

As mentioned earlier in this book, the National Interagency Fire Center monitors line-of-duty fatalities and injuries to wildland firefighters. The data has been maintained since 1910. As one would expect, not all injuries or fatalities have been reported—especially in the early years of organized wildland fire-fighting. The chart below depicts the line-of-duty injuries and/or fatalities (Top 8 categories) involving wildland firefighters. The information is provided in two categories: number of incidents (may or may not involve fatalities) and the number of fatalities. This information should be used as a visual reference of how many hazards or unique risks can impact our safety.

The National Interagency Fire Center (NIFC) identifies fatalities and injuries by type. For the purpose of this chapter, we have grouped accidents/fatalities in several larger groups. One example involves vehicle accidents or driving. NIFC lists 25 categories for driving related or vehicle accidents. For the purpose of this book, we will group all vehicle accidents or driving-related injuries/fatalities into one category. The same holds true for fixed-wing aircraft, burnovers, and medical related. The study provides a quick over-view of the accident types and does not examine all accidents or fatalities on an individual basis. Finally, the data provided in this study will encompass events that occurred between 1910 and 2013.

The following table further depicts the type of hazard that has been involved in firefighter accidents or fatalities. Again, some of the categories have been grouped together by hazard type and/or activity. The table should again serve as an illustration of the many types of hazards that can impact our safety on the fireline.

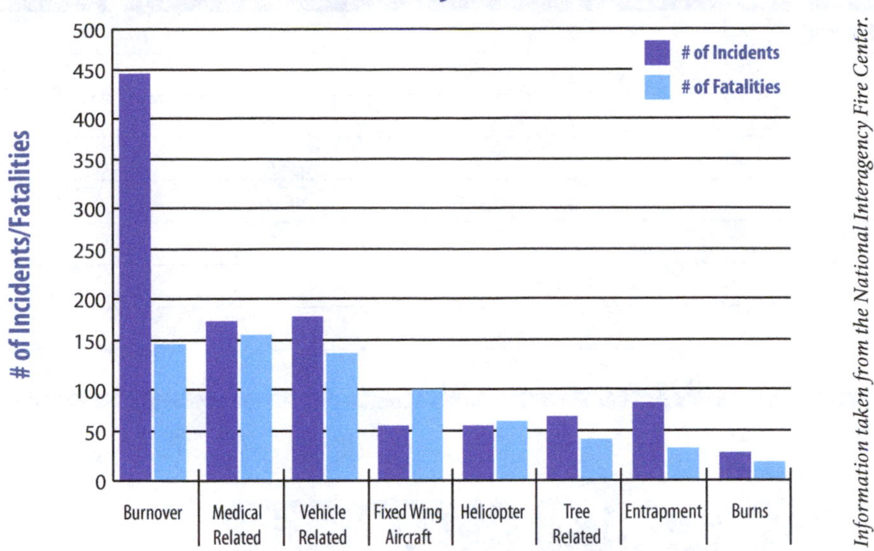

Figure 10-1 *Hazard Related Injuries and Fatalities*

TABLE 10-1 ***COMPLETE LIST OF HAZARDS ASSOCIATED WITH WILDLAND FIREFIGHTER INJURIES OR DEATHS***

Hazard Type	# of Incidents	# of Fatalities	Ranking #
Burnover	152	449	**1**
Medical Related	175	151	**2**
Vehicle Related	179	140	**3**
Fixed Wing Aircraft	54	99	**4**
Helicopter	55	56	**5**
Tree Related	68	44	**6**
Entrapment	86	37	**7**
Burns	33	21	**8**
Electrocution	12	11	**9**
Work Capacity Test	8	6	**10**
Asphyxiation or Suffocation	5	6	**10**
Heavy Equipment Related	11	5	**11**
Smokejumper Related	5	5	**11**
Drowning	4	4	**12**
ATV or UTV	5	3	**13**
Lightning	5	3	**13**
Rock (Rolling)	3	3	**13**
Fall	2	2	**14**

(continued)

Hazard Type	# of Incidents	# of Fatalities	Ranking #
Toxic Chemicals (Hazmat)	2	2	**14**
Hit by Train	2	2	**14**
Fire Training	2	2	**14**
Retardant Drop	2	2	**14**
Equipment Related	3	2	**14**
Hypothermia	1	1	**15**
Methane Gas	1	1	**15**
Murder	1	1	**15**
Flying Debris	1	1	**15**

SAFETY AND HAZARD CATEGORIES

Safety is the most important element of firefighting. Prioritizing safety in a dangerous and complex environment should be everybody's concern and responsibility. Safety is defined as: "The condition of being safe from undergoing or causing hurt, injury of loss."[1] Safety can also be described as the "freedom from exposure to danger, exemption from injury, and to protect from injury."[2] Safety may also involve a piece of personal protective equipment designed to protect the firefighter or a device designed to protect the user from harmful equipment (e.g. chainsaw chaps).

Wildland fire managers strive to create a cultural of safety. Fire mangers should strive to develop a safety culture and firefighters should ensure the cultural is maintained. The information below is provided by Michael J. Ward in his book, *Fire Officer: Principles and Practices, 3rd Edition* and describes a safety culture.

1. Provide honest sharing of safety information without the fear of reprisal from superiors.
2. Adopt a non-punitive policy toward errors.
3. Take action to reduce errors in the system. Walk the talk.
4. Train firefighters in error avoidance and detection.
5. Train fire officers in evaluating situations, reinforcing error avoidance, and managing the safety process.[3]

[1] National Wildfire Coordinating Group, *Firefighter Training, S-130* (Boise: National Wildfire Coordinating Group, 2003), 4D.3.

[2] National Wildfire Coordinating Group, *Training*, 2003. 4D.3.

[3] Michael J. Ward, *Fire Officer: Principles and Practices* (Burlington, Maine: Jones and Bartlett Learning, 2015), 387.

Fireline Hazards

Fireline hazards are classified into two distinctive categories: 1) Subjective and 2) Objective. Paul Gleason originally documented these two hazard categories in his creation of LCES. He explained subjective hazards as the hazards over which we have direct control. Objective hazards are the hazards over which we do not have direct control. They are part of the environment or fireline. The goal of every firefighter is to eliminate the subjective hazards or control them. Controlling the hazard is called **mitigation**.

Teaching Illustration: Look at a pocketknife. The cutting edge of the pocketknife is the objective hazard. We do not have direct control over the hazard unless mitigation measures are provided. If you run your hand over the blade (do not do this; just think about it) then you have subjected yourself to the objective hazard. As firefighters, our goal is to control or eliminate subjective hazards. If you close the knife blade, then you have mitigated the objective hazard.

If you encounter any hazard, regardless if it is subjective or objective, you must immediately notify your supervisor and other firefighters. Use flagging as a communication tool to identify the hazard. Do not assume that everybody will see the hazard. Make it safe and let people know! Do not continue working until the risk has been evaluated and mitigated. Mitigation measures can include complete avoidance. The key here is that the hazard must be identified and evaluated by the appropriate individuals before firefighters continue with their activities.

CATEGORIES OF FIRELINE HAZARDS

Fire Environment Hazards

Includes hazards that are common on the fireline.

Lightning  © Balazs Kovacs Images/ Shutterstock.com Danger: Lightning is extremely dangerous to firefighters.	**Fire Weakened Timber**  Courtesy of Brian Henington Danger: Trees become very weak and unpredictable.

(continued)

Rolling Rocks or Logs

© JaysonPhotography/Shutterstock.com

Danger: Being hit by the rolling rock or log.

Fire Entrapment or Burnover

© Ingrid Curry/Shutterstock.com

Granite Mountain Hotshot Memorial

Danger: Being overrun or entrapped by the fire.

Burns

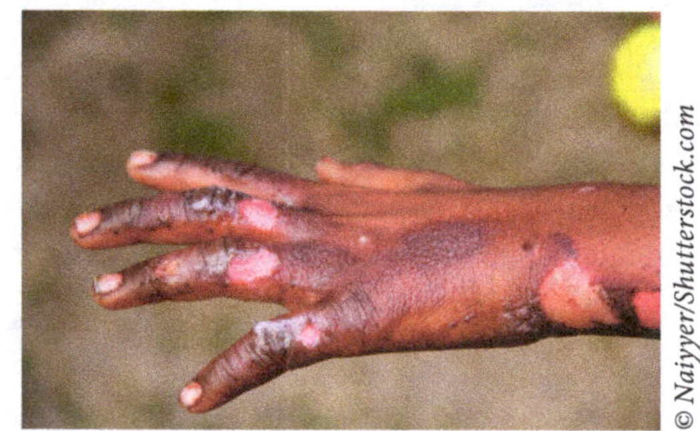

© Naiyyer/Shutterstock.com

Mitigation: Always wear your PPE.

Smoke

Courtesy of Brian Henington

1630 hours (4:30 PM). Complete smoke cover

Danger: Not able to see and impact your ability to function.

Falls/Steep Terrain

© Anton_Ivanov/Shutterstock.com

Danger: Slipping and falling.

Stump Holes

Courtesy of Brian Henington

Danger: Major source of firefighter burns.

Biological Hazards

Includes any living organism. Flora and fauna species are included in this category. In addition, viruses or bacteria are also included in this category. Depending on your jurisdictional responsibility, you may encounter specific biological hazards that can impact your safety. You should become familiar with these hazards and become familiar with proven mitigation techniques.

Snakes

© Audrey Snider-Bell / Shutterstock.com

Concerns: Poisonous and/or painful bites.

Wasps/Bees

© Renato Arap/Shutterstock.com

Concerns: Allergic reaction with some people. Swarming of wasps/bees may cause major issues.

Ticks

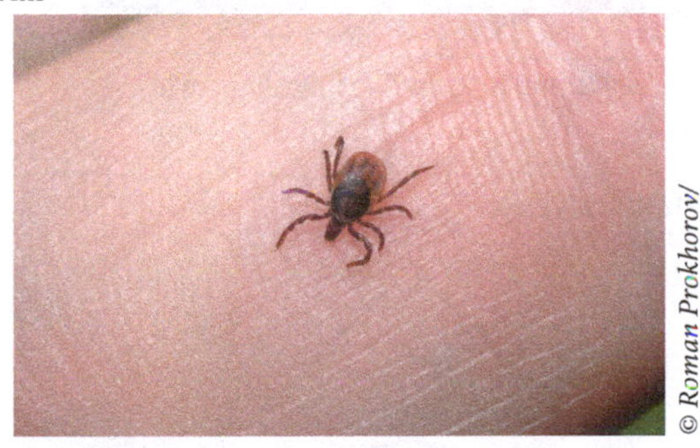

© Roman Prokhorov/ Shutterstock.com

Concerns: Carry diseases. Lyme disease and Rocky Mountain Spotted Tick fever, for example.

Scorpions

© Audrey Snider-Bell/ Shutterstock.com

Concerns: Some species are poisonous. Stings can be extremely painful.

Mosquitos or Gnats

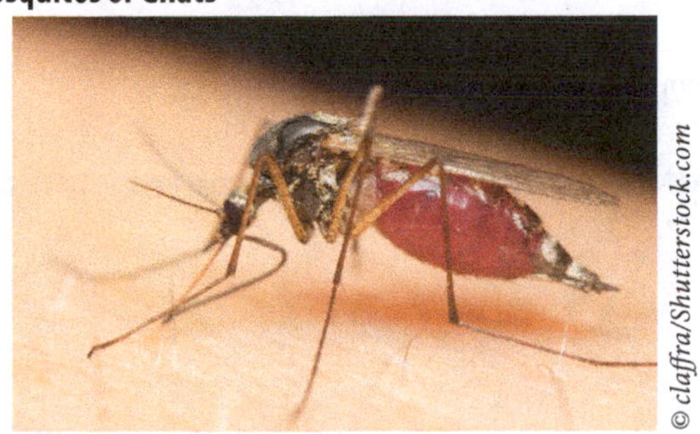

© claffra/Shutterstock.com

Concerns: Carry diseases. West Nile, for example.

Bears

© Derek R. Audette/Shutterstock.com

Concerns: Mother bear protecting her cubs. Bears seeking food from firefighters.

(continued)

Moose/Buffalo

Courtesy of Brian Henington

Concerns: Charging animals especially during the rut or breeding season.

Poison Oak or Poison Sumac

© Jerry-Rainey/Shutterstock.com

Concerns: Produces toxins when it is burned. Very painful sores if body becomes exposed.

Mountain Lions

© creativex/Shutterstock.com

Concerns: Fleeing animals.

Poison Ivy

© Tom Grundy/Shutterstock.com

Concerns: Painful sores if exposed. Produce toxins when burned by fire.

Marijuana

© David Maska/Shutterstock.com

Concerns: High levels of smoke will cause major medical issues. Booby traps should also be expected.

Viral Infections/Colds

©oculo/Shutterstock.com

Concerns: Sickness, camp crude.

Hazard Trees

Snags and/or hazard trees are extremely dangerous. A falling tree or debris falling from a tree can cause great harm or death to firefighters. As identified in Figure 10-1, snags, or falling debris from trees have resulted in 68 line-of-duty injuries and 44 firefighter fatalities.

Snags can be defined as: "A standing dead tree or part of a dead tree from which at least the leaves and smaller branches have fallen."[4] Snags may also include partially dying trees or even green trees if they are involved in active burning. Hazard trees should be elevated by trained professionals. Firefighters who are not trained should identify the hazard from a safe distance and immediately contact a supervisor. You should not allow anybody to enter the area and you should flag a perimeter around the snag.

The *Incident Response Pocket Guide* provides a clear set of instructions when dealing with hazard trees. It is highly recommended that you locate this guideline and become familiar with it.

There are visual indicators of snags or hazard trees. They may be very obvious or they may not be. Our heads should always be on a swivel. We should be constantly looking up to identify any potential overhead hazards. Some indications of hazard trees include:

- Dead standing trees
- Cat faces or fire scars
- Leaners (hang ups)
- Shallow roots
- Erosion near roots
- Specific tree species that present problems (white fir, cottonwoods, etc.)
- Widow makers (branches that are suspended in aerial fuels)
- Insect-infested trees
- Broken tops
- Mistletoe or conks

Snags or hazard trees are dangerous because it is hard to anticipate what they will or can do. A snag can fall without any warning. They can be impacted by heavy machinery or overhead aircraft. The wood is often weak or rotten. This creates complex issues as the direction of fall could be compromised. Wind also plays a major role as it can cause trees to fall or impact cutting operations. Below are some key considerations when dealing with hazard trees or snags.

Key Considerations:

- Only certified and qualified chainsaw operators can cut snags or hazard trees.
- If trees are being cut, all firefighters should be outside the cutting or drop zone of the tree. Stay out of the area.

[4] National Wildfire Coordination Group, *Glossary of Wildland Fire Terminology* (Boise: National Wildfire Coordinating Group, 2014), 161.

- Contact your supervisor immediately and stay out of the drop zone.
- Golden Rule: Flag a perimeter around the tree one and a half (1½) times the height of the tree. If the tree is 100 feet tall, then flag a complete circle around the tree at 150 feet. Don't assume that placing a piece of flagging will be appropriate in the identification of the hazard. Use your flagging. Make sure everybody is aware of the hazard tree.

Not every hazard tree or snag will be cut. Chainsaw operators evaluate the tree and determine if it is safe enough to bring down. If they determine it is not safe, then other measures can be taken to mitigate the risk. Some examples include:

- Using heavy equipment, such as a dozer or feller buncher.
- If the tree is already on fire, flag the drop zone and allow the tree to fall down on its own.
- Bucket drops from aircraft.
- Explosive crews.
- Dropping a solid and live tree into the hazard tree.

Most of these techniques are considered high risk. They should be evaluated by qualified personnel including safety officers. Remember, complete avoidance is acceptable and may be the best option for most hazard trees. If this option is selected, then flag a protection boundary around the tree to ensure other firefighters do not enter the area.

Visual Examples of Snags or Hazard Trees

(continued)

Leaning Trees (Leaners)

Courtesy of Brian Henington

Catface

Courtesy of Brian Henington

Down and Dead Trees

Courtesy of Brian Henington

Active Burning

Courtesy of Brian Henington

Rotten Trees

Courtesy of Brian Henington

High Density of Snags

Courtesy of Brian Henington

Human Activities

Human activities include any hazard that has been produced by humans. Examples include power lines, oil/gas activity, illicit labs, etc. Human activities can pose the most hazards to firefighters as they have unique and very hazardous conditions associated with them.

Power Lines

Courtesy of Brian Henington

Concerns: Electrocution is a very serious reality on the fire line. Evaluate your surroundings before you engage a fire. Power line Safety Mitigation measures are located in the *IRPG*.

Oil/Gas Activity

Courtesy of Brian Henington

Concerns: Deadly gases (H2S). Ruptured pipelines. Flammable and corrosive products. *IRPG* has mitigation measures when working in oil/gas fields.

Unexploded Ordnance

© *Dmytro Pylypenko /Shutterstock.com*

Concerns: More common than people think. *IRPG* has safety procedures for dealing with unexploded ordnance.

Illicit Labs

© *Mr. Green/Shutterstock.com*

Concerns: Booby traps, protection of site by producers. Meth labs are becoming more of an issue in recent years. HazMat scene.

Hazardous Materials

© *Mikadun/Shutterstock.com*

Concerns: Toxins, flammables, corrosives, etc. Avoid area, contact incident commander, and call for qualified and equipped HazMat technicians.

Propane Tanks

© *chloe7992/Shutterstock.com*

Concerns: High pressure, flammable gases. Rule of thumb: stay away. Propane tanks can BLEVE.

(continued)

Wildland/Urban Interface

© Darin Echelberger/Shutterstock.com

Concerns: Considerable amount of hazards. Will be covered in later chapters.

Panicked Public

© swatchandsoda/Shutterstock.com

Concerns: Evacuations, panicked public, high emotions, etc.

Illegal Dumps

© pixinoo/Shutterstock.com

Concerns: Toxins, chemicals, plastics, rubber, etc. Items should be treated as very harmful to wildland firefighters. Stay away. Secure scene, contact Incident Commander.

Abandoned Vehicles

Courtesy of Brian Henington

Concerns: Plastics, synthetic materials, rubbers, etc. Secure scene and contact appropriate authorities.

Working near the US/Mexico Border

© Frontpage/Shutterstock.com

Concerns: Drugs runners (mules), weapons, or illegal activity. You should be trained in border issues before engaging a fire along the border.

Firefighter Related

Attitude  *Courtesy of Brian Henington* Concerns: A positive attitude is conducive to safe and effective firefighting. A negative attitude is a major issue and should be addressed immediately.	**Physical Conditioning**  *Courtesy of Brian Henington* Concerns: Maintain the highest level of conditioning you can. Many injuries or deaths have occurred to firefighters who were not in the best conditioning.
Fatigue  *Courtesy of Brian Henington* Concerns: Distraction and lack of situational awareness. Not paying attention, negative attitude, etc.	**Experience Level**  *Courtesy of Brian Henington* Concerns: May impact a firefighter's ability to identify hazards and mitigate the hazards.
Training Level  *Courtesy of Brian Henington* Concerns: Improper training techniques and concepts can lead to serious injury or death.	**Critical Stress**  *© Patricia Marks/Shutterstock.com* Concerns: Firefighters are subject to situations that create critical stress. If you are involved in a situation—Get Help! There are numerous resources designed to help firefighters deal with critical stress.

Equipment Hazards

Hand Tools

Courtesy of Brian Henington

Concerns: Cover the cutting edges with a sheath or duct/filament tape.

Chainsaws

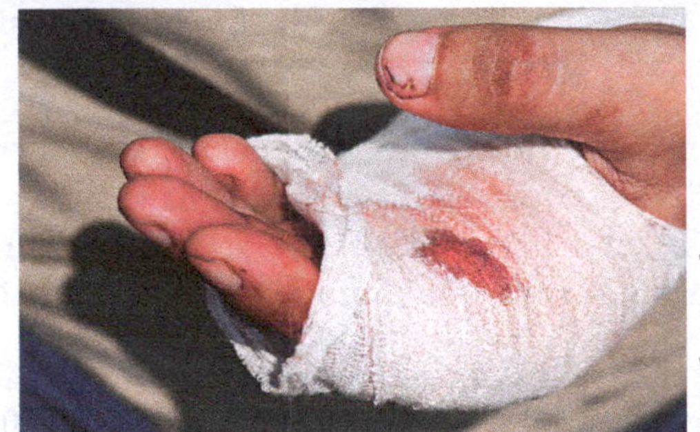

© Perfect Lazybones/Shutterstock.com

Concerns: Are very fast and sharp. Ensure you are wearing PPE and do not operate a chainsaw unless you are certified to do so.

Equipment Burns

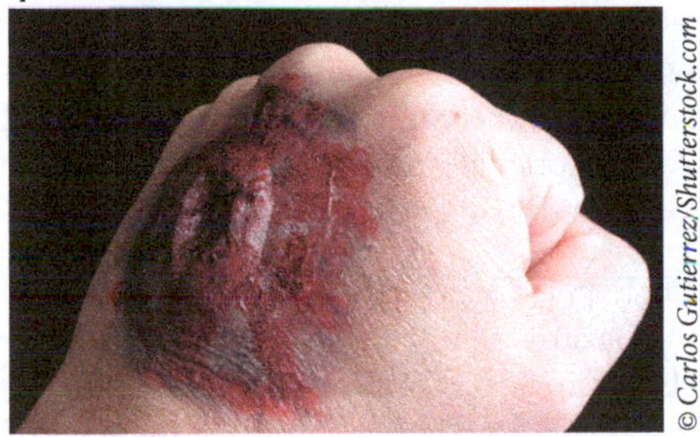

© Carlos Gutierrez/Shutterstock.com

Concerns: Wear your gloves at all times and be aware of your surroundings.

Noise

Courtesy of Brian Henington

Concerns: The fireline is saturated with loud equipment: pumps, chainsaws, engines, aircraft, heavy equipment, etc. Wear your ear protection.

Flying Debris/Wood Chips

Courtesy of Brian Henington

Concerns: Stay away from heavy equipment.

ATV/UTV

Courtesy of Brian Henington

Concerns: Very common tool used on the fireline. Not as stable as vehicles. Speed is a concern. Operators should be certified to operate an ATV/UTV and should wear the appropriate PPE.

Heartbreaking ATV Accident

In 2013, the New Mexico wildland fire community was devastated by the loss of a very respected and outstanding firefighter named Token Adams. Token was an Engine Captain on the Jemez Ranger District of the Santa Fe National Forest. On August 13, 2013, Token and his engine crew were trying to locate a small fire using their ATVs. The fire was located on Schoolhouse Mesa near the community of Jemez Springs, New Mexico.

In order to cover more ground, Token and the other members of his engine crew separated. They used their ATVs to cover a larger area and navigate a remote and densely forested area. The other two members of Token's crew were able to locate the fire around 1400 hours. They tried to contact Token on the radio to advise him of their findings. They were unable to do so.

Due to the inability to contact Token, a State Search and Rescue operation was initiated. Unfortunately and with great shock, Token's body was found a week after he went missing. He had died in the line of duty. The investigation of the accident discovered that Token's ATV had rolled over him, resulting in death.

Token had served his country with great distinction as a Navy veteran. He was also an experienced and qualified wildland firefighter which included being a former member of the Kings River Hotshots in California. He was a very skilled firefighter and certified to operate an ATV. Other firefighters remarked on how safely Token operated ATVs. "...one of the fire fighters commented on Token's conservative approach to ATV riding, mentioning how...he stuck to the roads and rode slow. If there were a lot of rocks, he'd walk."[5] He did not take chances or drive recklessly or fast. The investigators determined that Token's ATV hit a rock as he was trying to locate the fire. The rock caused the ATV to roll over on top of Token.

Token was very serious about safety. The accident was tragic! Experts believe the accident was not caused by Token's use of the ATV, but because of the environmental hazards that were present at the time of the fatality. This devastating accident should remind all of us how dangerous our job is. Token operated the ATV as his was trained, but the accident occurred anyway and New Mexico lost a valued member of the fire community and Token's wife and daughter are missing a husband and father.

[5] US Forest Service. *Schoolhouse Fire ATV Fatality: Learning Review Report* (Santa Fe: United States Forest Service, 2013), 5.

Vehicle Hazards

Fatigued Drivers

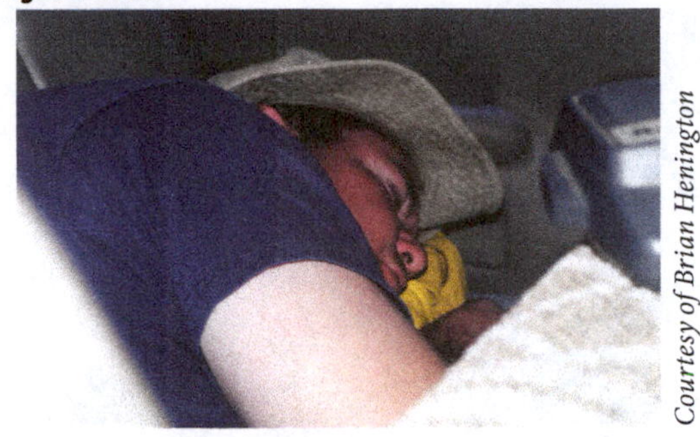

Courtesy of Brian Henington

Concerns: Fatigue is a major concern to firefighters. Tired drivers should rest before driving and adhere to the established driving standards and work/rest ratios.

Mechanical Failure

Courtesy of Brian Henington

Concerns: Mechanical failure of vehicles or equipment may cause accidents. The photo above shows a high-pressure oil line that ruptured on a crew carrier. Firefighters were able to catch the issue by conducting routine inspections of the vehicle.

Speed

Courtesy of Brian Henington

Concerns: Slow down!

Untrained Drivers

Courtesy of Brian Henington

Concerns: Fire vehicles may require specialize licensing or Commercial Driver's License. Do not operate a vehicle if you are not trained or qualified to do so.

Shifting Cargo

Courtesy of Brian Henington

Concerns: Secure cargo behind cage in enclosed vehicles. Tools should be fastened and not next to occupants (behind cargo cage).

Operating Near Personnel

© Andrew Orlemann/Shutterstock.com

Concerns: Operators may not be able to see firefighters on the ground.

Aircraft Hazards

Retardant Drops

© TFoxFoto/Shutterstock.com

Concerns: Fast moving aircraft dropping retardant on firefighters. Avoid working under aircraft.

Bucket Drops

Courtesy of Brian Henington

Concerns: 300 + gallons falling on you. Avoid working under aircraft.

Sling Loads

Courtesy of Brian Henington

Concerns: Static electricity and falling debris. Only trained personnel should support sling loads.

Rotor Wash

© You Touch Pix of EuToch/Shutterstock.com

Concerns: Turbulent winds can influence fire behavior or create very dusty conditions. Avoid working under aircraft.

Working Near Aircraft

© Stephanie Swartz/Shutterstock.com

Concerns: Potential of being hit by retardant or water drops. Avoid working under aircraft.

Falling Debris from Trees

Courtesy of Brian Henington

Concerns: Falling limbs or widow makers. Avoid working under aircraft.

Other Hazards

Heat-Related Illness

Courtesy of Brian Henington

Concerns: Can lead to dehydration, heat exhaustion, or heat stroke. Drink plenty of water.

Caves

Courtesy of Brian Henington

Concerns: Falling, serious injuries.

Hypothermia

Courtesy of Brian Henington

Concerns: Common during prescribed fire activities.

Darkness

Courtesy of Jim O'Leary IV

Concerns: Not able to adequately see hazards.

Carbon Monoxide

Courtesy of Brian Henington

Concerns: Odorless and tasteless. By-product of smoke. Shortness of breath, fatigue, headaches, fainting or worse.

Dust

© Tom Grundy/Shutterstock.com

Concerns: Impacts ability to see and operate vehicles safely. Also common around helispots and on dry dirt roads.

Carbon monoxide, carbon dioxide, and hydrogen cyanide are some by-products of fire and are found in smoke. They can all be extremely dangerous to firefighters and cause major medical issues. Since 1910, six wildland firefighters have died due to asphyxiation or suffocation.[6] Tactical activities should consider firefighter exposure to smoke and strive to limit the direct impact smoke has on firefighters. A bandana is not going to protect your airways in an effective manner. Step out of the smoke. The chart below is provided by the *WFSTAR: Wildland Fire Safety Refresher* series. It identifies the medical dangers associated with prolonged exposure to smoke.

TABLE 10-2 ***MEDICAL ISSUES RELATED TO PROLONGED EXPOSURE TO SMOKE***

CO in Atmosphere (ppm)	COHb in Blood (%)	Signs and Symptoms
10	2	Asymptomatic
70	10	No appreciable effect, except shortness of breath on vigorous activities; possible tightness across the forehead.
120	20	Shortness of breath on moderate exertion; occasional headache with throbbing in temples.
220	30	Headache; irritable; easily fatigued; judgment disturbed; possible dizziness; dimness of vision.
350–520	40–50	Headache, confusion; collapse; fainting on exertion.
800–1220	60–70	Unconsciousness; intermittent convulsion; respiratory failure, death if exposure is long continued.
1950	80	Rapidly fatal

Information taken from WFSTAR: Annual Wildland Fire Safety Refresher Training: National Interagency Fire Center.

[6] Information gathered from *National Interagency Fire Center.* 2015.

Base/Camp Hazards

Includes hazards located at the Incident Base and/or Camp.

Poor Lighting

© Kzenon/Shutterstock.com

Concerns: Not able to see.

Sleeping Areas

© TFoxFoto/Shutterstock.com

Concerns: Sleeping areas to close to vehicles or helibase. Biological hazards may also be present (i.e., snakes).

Sanitation

© Leena Robinson/Shutterstock.com

Concerns: Transfer of germs and other hazardous biological elements.

Extension Cords

© Greg Henry/Shutterstock.com

Concerns: Tripping and falling. Usually found around base or camp facilities.

Steep Steps

© Yuttasak Jannarong/Shutterstock.com

Concerns: Slipping and falling. Steep steps are usually associated with food catering services or loading onto certain crew transport vehicles.

Handling of Food

© Poznyakov/Shutterstock.com

Concerns: Transfer of germs, bacteria, etc.

FIREFIGHTERS' RIGHT TO REFUSE RISK

There may be certain times in your career that require you to refuse an assignment or task. This concept is called refusing risk. The decision to refuse risk is based on safety concerns. As firefighters, we have the right to refuse unsafe work tasks or assignments, and in fact, we have an obligation to do so. According to the *Incident Response Pocket Guide*, "When an individual feels an assignment is unsafe they also have the obligation to identify, to the degree possible, safe alternatives for completing that assignment. Turning down an assignment is one possible outcome of managing risk."[7] Supervisors and fire managers should be considerate of the turn down and reevaluate the situation to ensure proper mitigation procedures are identified and put in place.

The four factors involved with refusing risk (or turn down) are:

1. There is a violation of safe work practices.
2. Environmental conditions make the work unsafe.
3. They lack the necessary qualifications or experience.
4. Defective equipment is being used.[8]

SUMMARY

The wildland firefighter will be exposed to numerous fireline hazards. Successful and safe firefighters are able to identify the risk associated with the hazards and identify appropriate mitigation measures to rectify the situation. As firefighters, we must know how these risks can impact our safety and the safety of other firefighters. When you encounter a fireline hazard, stop your activities, notify your supervisor, and identify the hazard through flagging so other firefighters are also aware of the hazard. Do not continue activities until a qualified firefighter has effectively evaluated the risk and the exposure to firefighters.

[7] Incident Response Pocket Guide, 2014, 19.

[8] Ibid.

KNOWLEDGE ASSESSMENT

1. What is the number one cause of wildland firefighter fatalities? How may fatalities have been involved with helicopters? How many fatalities from electrocution?
2. What is the definition of safety?
3. What are the five (5) steps involved with a safety culture?
4. What are the two categories of fireline hazards? Who created these categories?
5. Which hazard category is controllable?
6. Controlling the hazard is called ________________________?
7. What communication device can you use to identify hazards?
8. What is the definition of a snag?
9. What are three byproducts (chemicals) found in smoke?
10. Does the firefighter have the right to refuse risk? What is it based on?

EXERCISES

1. Study Table 10-1. What hazard type related to fatalities stood out to you? Why?
2. Locate Thunderstorm Safety in the *Incident Response Pocket Guide.* How many steps are involved in the process? In addition, how many firefighters have died in the line of duty due to lightning strikes?
3. Locate Hazard Tree Safety in the *Incident Response Pocket Guide.* How many steps are involved in the process? In addition, how many firefighters have died in the line of duty due to tree related activities?
4. Located How to Properly Refuse Risk in the *Incident Response Pocket Guide.* What are the four factors involved with refusing risk.

BIBLIOGRAPHY

National Wildfire Coordinating Group. Firefighter *Training, S-130.* Boise: National Wildfire Coordinating Group, 2003.

National Wildfire Coordinating Group. *Incident Response Pocket Guide.* Boise: National Wildfire Coordinating Group, 2014.

National Wildfire Coordinating Group. *Glossary of Wildland Fire Terminology.* Boise: National Wildfire Coordinating Group, 2014.

National Interagency Fire Center. *WFSTAR: Wildland Fire Safety Refresher. Smoke: Knowing the Risks. Instructor Manual.* Boise: National Interagency Fire Center, 2014.

United States Forest Service. Santa Fe National Forest. *Schoolhouse Fire ATV Fatality: Learning Review Report.* Santa Fe: United States Forest Service, 2013.

Ward, Michael J. *Fire Officer: Principles and Practice. 3rd Edition.* Burlington, MA: Jones and Bartlett Learning, 2015.

CHAPTER 11

SITUATIONAL AWARENESS AND THE RISK MANAGEMENT PROCESS

Learning Outcomes

- Explain the concept of situational awareness and how it relates to safe firefighting.
- Describe the five communication responsibilities used during fire suppression activities.
- Effectively explain the risk management process and how it is used in fireline decision making.
- Recognize the barriers to effective situational awareness.

Courtesy of Mark Meyers

Key Terms

After Action Review
Barriers to Situational Awareness
Communication Responsibilities
Direct Communication
Hazardous Attitudes
Situational Awareness
Standard Operating Procedures
Risk Management Process

OVERVIEW

The last few chapters have concentrated on the critical importance of the Rules of Engagement as well as introducing you to fireline hazards. It should be obvious how dangerous and complex our profession is. So how do you stay safe while fighting a fire? This concept is explained as the **Risk Management Process**. The foundation of this proven and effective system is based on scientific research on the human's ability to process critical and vital information and make appropriate and safe decisions. The center focus of this process is a concept called situational awareness.

This chapter will introduce situational awareness as well as concentrating on the importance of communication to safe firefighting activities. In addition, you will also be introduced to the five steps involved in the risk management process. The intent of this chapter is to introduce the entry-level firefighter to the core components needed to develop a functioning and effective risk management system.

SITUATIONAL AWARENESS

Dynamic, complex, and dangerous professions have utilized a proven process to effectively evaluate their surroundings and make appropriate decisions based on their observations and experience. High risk or high reliability organizations must have an understanding of the situation they are involved in and how to make appropriate and safe decisions. This system is called **Situational Awareness**. Situational awareness is explained by the National Wildfire Coordination Group as, "An on-going process of gathering information by observation and by communication with others. This information is integrated to create an individual's perception of a given situation."[1] The process can be further explained as "An individual's perception and comprehension of the details of their surrounding environment, and the understanding of how events occurring in the moment may affect the future."[2]

The goal of every firefighter should be to develop a functioning and effective situational awareness process. You will begin to develop this process by applying the skills and knowledge you learn in class. You will further develop this process by enhancing your skills through fireline experience and continued training.

The ability to maintain an effective and functioning situational awareness may be the most challenging component of emergency response activities. As you have learned in previous chapters, the complex field of wildland fire is always changing. Wind, terrain, and fuel conditions can impact your current tactical application and ultimately change the situation. You should always update the situational awareness cycle by constantly evaluating your surroundings and using the Rules of Engagement to determine the best solution. Furthermore, the ability to identify and analyze fireline hazards also plays an essential role in a functioning awareness cycle.

---

[1] National Wildfire Coordinating Group, *Glossary of Wildland Fire Terminology* (Boise: National Wildfire Coordinating Group, 2014), 159.

[2] Leslie Miller and Clint Clausing, International Fire Service Training Association, *Hazardous Materials for First Responders, 4th Edition* (Stillwater, Oklahoma State University: Board of Regents, 2010), 275.

An effective situational awareness cycle must include or incorporate the following:

1. Always functioning: The cycle only works if it is constantly being updated and reevaluated.
2. Evolve and Adapt: The cycle has to evolve with the events of the day to be effective. A good situational awareness cycle helps us anticipate changes.
3. Gather Information: Requires constant information gathering. This can be done through personal observations or the observations of others. Good decisions rely on good information.
4. Communicate: Communication has to be maintained and prioritized.
5. Top Priority: Considered the most important step in the risk management process.

It is easy to become distracted on the fireline. Distractions can come in many forms: fatigue, attitude, fire behavior, stress, media, etc. In order to facilitate an effective situational awareness cycle, you must recognize the common distractions that impact the cycle and how to effectively deal with these distractions. Below is a list of some distractions that may impact a functioning awareness cycle:

- Inexperience
- Stress
- Fire Behavior
- Fatigue
- Attitude
- Information Overload
- Tunnel Vision
- Wildland/Urban Interface
- Hazardous Materials
- Media
- Politicians
- Panicked Public

A very common distraction on the fireline is to become focused on one activity. Often called tunnel vision, this fixation does not allow firefighters to see the entire picture. Two psychology professors from Harvard University studied this occurrence as it relates to dynamic events. Daniel J. Simons and Christopher F. Charbris called this psychological phenomenon inattentional or change blindness. This concept is explained best in firefighting terms as not effectively viewing your entire picture or becoming fixated on one activity. This is very dangerous in emergency response activities because it does not allow the firefighter to witness changes and apply the appropriate safety precaution.

The professors explain that "Studies of intentional blindness have made an even stronger claim: that, without attention, visual features of our environment are not perceived at all (or at least not consciously perceived)—observers may fail not just to change detection, but perception as well."[3] The professors conducted their studies through a series of scientific observations that were documented on videos. One video in particular shows a group of people wearing black and white shirts. The video asks the viewer to count how many times the group wearing white shirts passes a basketball. As the viewer becomes focused on the task at hand, the video introduces a gorilla. Most viewers do not see the gorilla because they become fixated on the task at hand; they don't see the entire picture.

As emergency responders, we have to notice subtle and small changes in our environment. We cannot become focused solely on the task at hand. Your ability to notice changes during fire activities is vital. Using the Rules of Engagement and the concepts included in Look Up, Look Down, and Look Around will help you to monitor changing dynamics and react accordingly.

[3] Daniel J. Simons and Christopher F. Chabris, "Gorillas in Our Midst: Sustained Inattentional Blindness for Dynamic Events," *Perception*, Volume 28 (1999): 1060.

COMMUNICATION RESPONSIBILITIES

The importance of communication during fire activities is essential to effective and safe firefighting. Without adequate communications, the Rules of Engagement are not in place and, in fact, are violated. This section will focus on the concept of communications as it relates to emergency response activities. Situational awareness relies on good and effective communication. Without proper communication, our ability to make decisions is compromised.

WHY IS COMMUNICATION IMPORTANT TO FIREFIGHTERS?

1. Communication is required for you to receive orders and instructions.
2. Communication is required for you to be aware of many new hazards and safety concerns.
3. Communication is required for you to make others aware of what you see in a situation.
4. Communication will be how you will learn about the business of firefighting.[4]

Sender/Receiver Communication

In communication, the sender is the person who initiates the communication. An example of this is your instructor in a classroom. The instructor will begin the transfer of information and continue passing the information through the lecture. The receiver is the person for whom the communication is intended. In a classroom, the students are the receivers. On a fireline, the sender/receiver may be anybody. The process does not necessarily include a supervisor directing the employees. It can include two firefighters who encounter a hazard. As one begins the conversation, he or she is considered the sender. The receiver listens to the sender and then provides feedback.

The communication technique described above can be explained as feed-forward and feedback. The process of feed-forward/feedback is important because it allows for clarification and questions. It fulfills the Rules of Engagement by ensuring that instructions are given and understood. The figure below depicts the relationship between a sender and receiver.

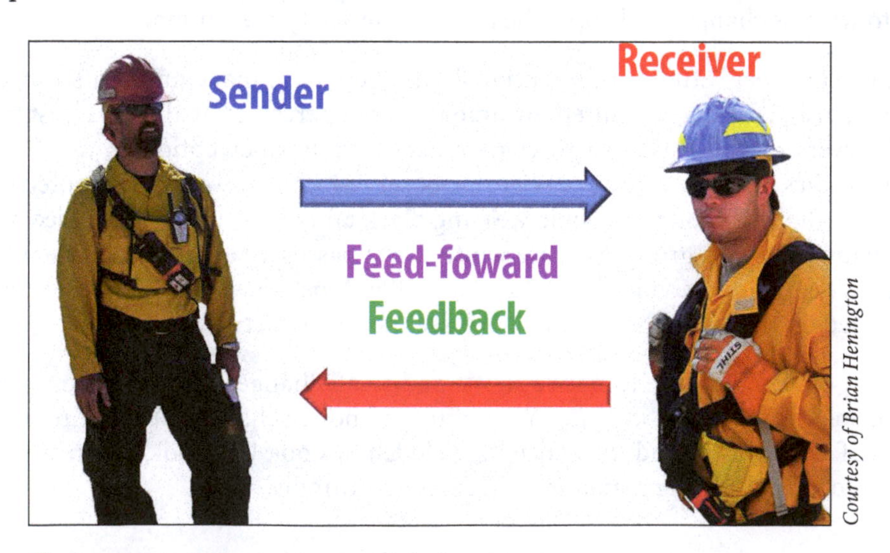

Figure 11-1 *Communication Relationship.*

[4] National Wildfire Coordinating Group, *Firefighter Training, S-130* (Boise: National Wildfire Coordinating Group, 2003), 4D.7.

There are some issues with receiver and sender communication. The common goal with this process is to ensure that you ask questions if you do not understand; if you are giving the information (acting as the sender), make sure you allow for questions. The information below depicts common receiver and sender errors.

TABLE 11-1 *COMMON SENDER AND RECEIVER ERRORS*[5]

Common Sender Errors	Common Receiver Errors
• Not establishing a common frame of reference	• Listening while locked into the context of a previous perception
• Omission of information	• Not being prepared
• Providing biased or weighted information	• Thinking ahead of the speaker
• Assuming the message depends only on the words used	• Inattention to non-verbal signals
• Not willing to repeat information	• Not asking for clarification
• Disrespectful communication	• Disrespectful communication

The lack of quality and effective communication is a major concern on a fireline. To improve this issue, we must realize what our communication issues are. Do you avoid talking when you get tired or hungry? Do you dislike the way another agency was treating you? These are just examples of situations that may impact effective communication. I always tell my firefighters that they should be exhausted by the end of the day because of their constant communication. We must continually communicate our observations. A major mistake is to assume that all firefighters are seeing what you are witnessing. Communicate the issue and ensure your receiver is comprehending the message.

Direct Communication

Direct communication is the preferred communication technique on emergency response activities. Direct communication is described as straightforward and precise. Direct communication should not include long, confusing statements. The proper use of direct communication will expedite the communication process and facilitate the effective transfer of information. Direct statements are the recommended communication technique when talking on a radio.

The information below will help you develop effective direct communication techniques.

1. Be respectful. Direct communication can often be viewed as assertive or rude. Make certain you are respectful in your transfer of the information to ensure the receiver of the information does not check out or quit listening to you.
2. Use the person's name or call sign. This will immediately alert the receiver of the need to listen.
3. Use the word "I." This will establish the sender as the person initiating the communication.
4. Be clear and precise in the delivery of your message.
5. Use appropriate emotion. This will relay to the sender the urgency of your message. This concept should not be confused with acting emotionally. Emotion helps us to get the message across. An example of what should not be done is yelling, screaming, or cussing at somebody. This breaks down the communication process. An example of appropriate emotion is: Direct, precise communication with a calm and collective demeanor.

[5] National Wildfire Coordinating Group. L-180 Human Factors on the Fireline. 2008.

6. Employ feedback. Ask a question to the receiver such as "Do you understand what I want?" This technique will ensure that the receiver has time to clarify the message.
7. Do not disengage the conversation until both parties are clear on the message.

Example of direct communication between a division supervisor and an engine.

Division Supervisor:	Engine 42, Division Alpha.
Engine 42:	This is Engine 42, go ahead Division Alpha.
Division Supervisor:	Engine 42, I need you to relocate to Helispot 3. We have a spot fire and need your assistance immediately!
Engine 42:	Engine 42 copies, enroute to Helispot 3 to assist with a spot fire. ETA 10 minutes.
Division Supervisor:	Copy Engine 42. Do you have any questions?
Engine 42:	No, Division Alpha. Engine 42 clear.
Division Supervisor:	Division Alpha clear.

The example above illustrates the components for effective direct communication. Division Alpha was precise with the communication and stressed the urgency for Engine 42's assistance. He was direct, but not disrespectful, nor did he shout or yell the message. Engine 42 was also precise in its response and confirmed the message received by stating the estimated time of arrival. Division Alpha also allowed for the engine to ask questions or clarification.

Communication Responsibilities

Communication responsibilities should be a priority for every firefighter. It is our responsibility to use these guidelines to facilitate effective communication. In my experience, the number one complaint by firefighters is that their boss will not communicate with them or lacks effective communication skills. If we all use the techniques described here, we will ensure that communication is prioritized and maintained at the highest and most effective manner possible.

The five communication responsibilities are identified below:

1. Brief others: Provide an adequate and effective briefing. This does not include only supervisors. This can include a firefighter briefing another firefighter on a hazard. Use standard operating procedures to help with this process. The *Incident Response Pocket Guide* provides a briefing checklist that will help you cover the important components of a briefing.
2. Debrief your actions: After Action Reviews are essential to effective firefighting. The concept will be explained later in the chapter.
3. Communicate hazards to others: Do not assume everybody is seeing what you are. Identify hazards with flagging.

4. Acknowledge messages and understand intent: Do you understand your leader's intent or assignment? If you do not, ask for clarification. On the other hand, fire managers must allow for clarification.
5. Ask if you do not know: Do not begin or continue activities if you do not understand what needs to happen.

Standard Operating Procedures (SOPs) are guidelines used by emergency responders to facilitate actions and communications. The National Wildfire Coordination Group clarifies standard operating procedures as "Specific instructions clearly spelling out what is expected of an individual every time they perform a given task. A standard operating procedure can be used as a performance standard for tasks that are routinely done in the operational environment."[6] Example of SOPs used on wildland fire include:

- Standard fire orders
- Dozer hand signals
- Helicopter hand signals
- Radio call signs
- Dispatch protocol
- Radio frequencies

After Action Review (AAR) is a highly effective method used to debrief the actions of a day. After Action Reviews are described as:

> *A structured review or de-brief process of an event, focused on performance standards, that enables participants to discover for themselves what happened, why it happened, and how to sustain strengths and improve on weaknesses. After Action Reviews, informal or formal, follow the same general format, involve the exchange of ideas and observations, and focus on improving performance.*[7]

The *Incident Response Pocket Guide* provides a guideline for conducting an AAR. Most fire managers conduct AARs at the end of the operational period. A recent trend is to conduct AARs during tactical pauses or during lunch. The problem with conducting them at the end of the operational period is that if the activities end at 0200 hours, the process may not be very effective because firefighters are tired and want to sleep.

The four questions to be asked during an AAR include:

1. What was planned?
2. What actually happened?
3. Why did it happen?
4. What can we do better next time?

An AAR is effective only if all members are involved. Everybody should participate for the process to be an effective learning mechanism. The process should always be constructive, and should NOT provide an opportunity to vent your frustrations or anger.

---

[6] National Wildfire Coordinating Group, *Glossary*, 165.

[7] Ibid., 23.

RISK MANAGEMENT PROCESS

The risk management process is essential to safe firefighting activities. This process allows firefighters to gauge and plan for changes in the fire environment and addresses concerns or hazards that threaten firefighters. This process is critical in the preplanning process used by fire managers, but should also be used by every firefighter to gauge concerns and effectively address them. One of the most critical components involved in this process is time. Fire behavior and hazards may change throughout the day. The risk management process should always be functioning and reevaluated throughout the day to identify and manage risks. The figure below depicts the risk management process.

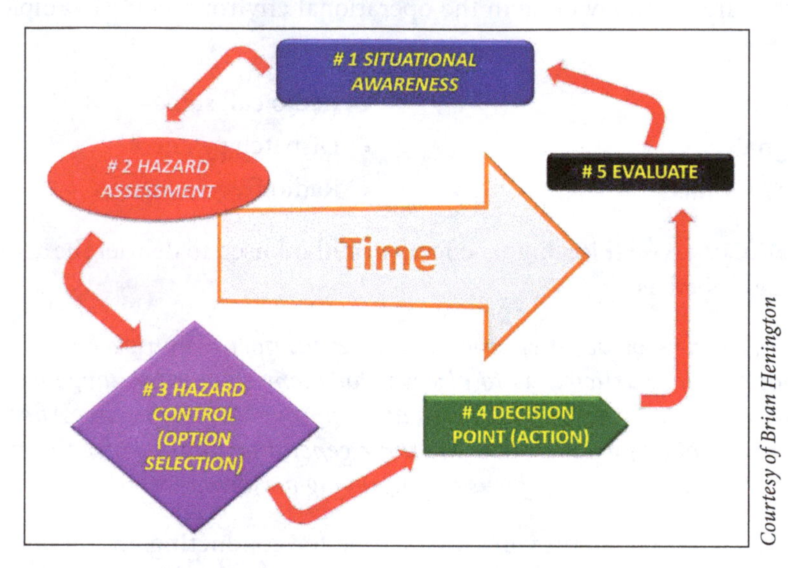

Courtesy of Brian Henington

Figure 11-2 *Risk Management Process.*

Step 1: Situational Awareness

A functioning and effective situational awareness process allows firefighters to witness the conditions and predict changes to the fireline environment. Maintaining effective communication and updating information will help develop this process. This step allows us to base our actions on the reality present on the fireline.

Step 2: Hazard Assessment

This step determines the hazards that present danger to firefighters. It includes recognizing risks and fireline hazards. Hazard assessment utilizes the Watch Out Situations and revolves around the concepts included in Look Up, Look Down, Look Around and Common Denominators of Fire Fatalities.

[8] Ibid., 22.

Step 3: Hazard Control

This step involves the decision to mitigate the hazard or risk. Hazard control "establishes the best way to minimize or control the risk from the hazards you will encounter while getting the job done."[8] Hazard control utilizes LCES and the Standard Firefighting Orders as the primary engagement tool for the analysis of tactical decisions. Finally, this step includes the option selection chosen to mitigate the hazard(s).

Step 4: Decision Point

The hazard or risk has been identified and the selection of mitigation has been chosen. The decision point is the actions we take to mitigate the risk/hazard. This process can include a Go/No Go Checklist. You should be aware of the decision and ensure your safety will not be compromised. Some decisions may be high-level management decisions (how to deal with extensive snags) or may be a crew level decision (where your lookout will be posted) or may be an individual decision (how to avoid a rattlesnake).

Step 5: Evaluate

Evaluate the actions selected to mitigate the risk/hazard. Was it effective? Did it compromise the safety of firefighters? The evaluation step should constantly be evaluated and readdressed as the events of the day progress. The evaluation step can be included in the After Action Review.

THE DECISIONS FIREFIGHTERS MAKE (THE HUMAN FACTOR)

A firefighter has to constantly make decisions that impact safety. Are fires or hazards killing firefighters or are the decisions firefighters make truly at the center of these tragedies? "Most fatalities and accidents are the direct result of errors in human judgment."[9] This assumption is what fire mangers refer to as human factors on the fireline. The National Wildfire Coordination Group has designed training curriculum to specifically address this issue. We have to understand that our decisions—or the lack of—are what we must control. Bad decisions can lead to fire-related injuries or fatalities.

Developing a functioning situation awareness process is the first step to effective risk management; however, the awareness cycle may be negatively impacted by several factors. We must understand what these issues are so we can identify and address them.

The first barrier to situational awareness is stress. The Stress Management Society says "Stress happens when we feel that we can't cope with pressure and this pressure comes in many shapes and forms, and triggers physiological responses."[10] Stress can have major emotional impacts on firefighters. They

[9] Ibid., 4D.4.

[10] Stress Management Society, *What is Stress?*, accessed March 7, 2015, http://www.stress.org.uk/what-is-stress.aspx.

can include elevated heart rate, headaches, frustration, anger, fatigue, and/or depression. The *Incident Response Pocket Guide* identifies stress reactions as it relates to safe firefighting activities. They are summarized below:

- Communication deteriorates or grows tense.
- Habitual or repetitive behaviors.
- Target fixation—Locking into a course of action, whether it makes sense or not and assuming attitude of "just try harder."
- Action tunneling—Focusing on small tasks, but ignoring the big picture.
- Escalation of commitment—Accepting increased risk as completion of task gets near."[11]

In addition to stress, there are several other factors that impact your ability to effectively analyze your surroundings. Some of the barriers are identified in the table below.

TABLE 11-2 ***BARRIERS TO SITUATIONAL AWARENESS***[12]

Low Experience Level with Local Factors	• Unfamiliar with the area or the organizational structure
Distraction from Primary Task	• Radio traffic • Conflict • Previous errors • Collateral duties • Incident within an incident
Fatigue	• Carbon Monoxide • Dehydration • Heat stress • Poor fitness level reduces resistance to fatigue • 24-hours awake affects your decision making capability like .10 blood alcohol content.

Finally, another obstruction to situational awareness is **hazardous attitudes**. Have you ever worked with a negative person? It's tough and exhausting, isn't it? Hazardous attitudes can directly and negatively impact a safe working environment. Hazardous attitudes are classified into several categories and must be addressed immediately.

The first step is to determine if your personal attitude is the issue. We should look at ourselves first to see if we are the problem. If we are, we need to fix our attitude because our behavior can impact effective communications, our safety, and more importantly, the safety of our fellow firefighters.

---

[11] National Wildfire Coordinating Group, *Incident Response Pocket Guide* (Boise: National Wildfire Coordinating Group, 2014), XI.

[12] National Wildfire Coordinating Group, *Incident Response Pocket Guide* (Boise: National Wildfire Coordinating Group, 2014), X.

The National Wildfire Coordinating Group identifies several hazardous attitudes as they relate to wildland firefighting operations. They are identified in the table below.

TABLE 11-3 *HAZARDOUS ATTITUDES*[13]

Type of Hazardous Attitude	Explanation
Invulnerable	That can't happen to us
Anti-authority	Disregard of the team effort
Impulsive	Do something even if it's wrong
Macho	Trying to impress or prove something
Complacent	Just another routine fire
Resigned	We can't make a difference
Group Think	Afraid to speak up or disagree
Mission Focused	Too engaged in accomplishing the task
Bias	Sexist, racist, sexual orientation, agency affiliation

Barriers to situational awareness can negatively impact your ability to effectively analyze and evaluate your surroundings. You cannot have an effective situational awareness process if you do not address and recognize the barriers to this system.

SUMMARY

Situation awareness is the most important component of the risk management process. Your ability to effectively evaluate your surroundings and make the appropriate and safe decision is critical to your success as a firefighter. Understand that barriers exist and how they can impact your situational awareness. Experience will help you find techniques to manage these barriers. In absence of fireline experience, try using techniques to identify barriers in your everyday life. Find proven techniques to deal with stress and hazardous attitudes.

[13] National Wildfire Coordinating Group, *Incident Response Pocket Guide* (Boise: National Wildfire Coordinating Group, 2014), XI.

KNOWLEDGE ASSESSMENT

1. What is situational awareness?
2. Identify the five steps involved in situational awareness.
3. Who is the sender? Who is the receiver?
4. Explain direct communication.
5. What are the five common responsibilities of communication?
6. What is the definition of standard operating procedures? Give five examples of SOPs used during wildland fire activities.
7. What are the four questions involved in an AAR?
8. The AAR is only effective if all members are ________________.
9. What are the five steps in the decision making process?
10. What are the hazardous attitudes identified in this chapter?

EXERCISES

1. Identify a time in your career or in college that you were overwhelmed with stress. How did the stress impact you? What did you do to deal with the stress? Can these techniques be used on the fireline? If not, have you explored ways to deal with stress in a productive manner?
2. Develop a personal safety culture. If human error is at the root of most fatalities or injuries, then how do you propose to personally address your ability to make the proper and correct decisions to eliminate human error?
3. Have you personally witnessed a hazardous attitude in your career or in school? What was the attitude? Was the issue addressed and if so, how was it addressed? Was it effective?

BIBLIOGRAPHY

Miller, Leslie and Clint Clausing. International Fire Service Training Association. *Hazardous Materials for First Responders, 4th Edition*. Stillwater, Oklahoma State University: Board of Regents, 2010.

National Wildfire Coordinating Group. *Firefighter Training, S-130*. Boise: National Wildfire Coordinating Group, 2003.

National Wildfire Coordinating Group. *Incident Response Pocket Guide*. Boise: National Wildfire Coordinating Group, 2014.

National Wildfire Coordinating Group. *Glossary of Wildland Fire Terminology*. Boise: National Wildfire Coordinating Group, 2014.

National Wildfire Coordinating Group. *Human Factors in the Wildland Fire Service, L-180*. Boise: National Wildfire Coordinating Group, 2008.

Simons, Daniel J. and Christopher F. Chabris. "Gorillas in Our Midst: Sustained Inattentional Blindness for Dynamic Events." Perception, Volume 28 (1999): 1059–1074.

Stress Management Society. *What is Stress?* http://www.stress.org.uk/what-is-stress.aspx.

CHAPTER 12

COMMON HAND TOOLS AND FIRING DEVICES

Learning Outcomes

- Identify the common hand tools used on wildland fires.
- State the importance of inspecting a hand tool prior to use on the fireline.
- Demonstrate the safety measures used when carrying tools and working next to other firefighters.
- Explain the importance of a tool lay and pack lay.
- Identify the two common firing devices used on wildland fires and explain the safety considerations for each.

Courtesy of Brian Henington

Key Terms

Adz Hoe
Combination Tool
Drip Torch
Firing Devices
Fire Shovel
Fusee
McLeod
Pulaski
Rhino Tool
Swatter/Flapper
Tool Lay and Pack Lay

OVERVIEW

Your hand tool is your livelihood. Your tool is how you make your money. You should treat your tool with the utmost care and respect. The better your tool is maintained, the easier your job will be.

This chapter will introduce you to the three categories of hand tools used on wildland fires. Firefighters and merchants of fire equipment make a wide variety of specialized tools. Some tools are highly effective, but for the purpose of time, we will introduce you only to the common hand tools used on fires.

In addition, fighting fire with fire is one of the preferred methods for suppressing wildland fires. This chapter will also introduce you to the common firing devices used by wildland firefighters. You will be introduced to the fusee and the drip torch. The safety precautions for these devices will also be covered.

Courtesy of Brian Henington

COMMON WILDLAND FIRE HAND TOOLS

Wildland fire hand tools are classified into three categories. They include: cutting tools, scraping tools, and smothering tools. Your choice of tool will depend on the activities you are engaged in or depending on your role on your crew. Most supervisors will assign tools to each firefighter and their decision is usually based on the work ethic or strength of each crew member.

Cutting Tools

Cutting tools are very popular among firefighters. They are used to construct fireline, cut brush or small trees, and/or for grubbing and trenching. The most common cutting tool is the Pulaski (named after the iconic wildland firefighter Edward Pulaski). Other cutting tools include single-bit ax, adz hoe, or other specialized tools designed to cut or construct fireline.

Pulaski

- Best Uses/Characteristics
  - » Provides cutting and grubbing edge
  - » Great for fireline construction
  - » Cuts small trees and brush
- How to Use
  - » Swing at 45° angle
  - » Hold securely
  - » Feet should be shoulder width apart
  - » Stand on firm ground
- Disadvantage
  - » Hard on the firefighter's back
- Sharpening
  - » Taper cutting edge 2″ wide with even bevel
  - » Bevel grub edge 3/8″ wide on a 45° angle

Courtesy of Brian Henington

Single-bit Ax

- Best Uses/Characteristics
  - » Supports chainsaw operations
  - » Driving wedges
- How to Use
  - » Hold securely
  - » Feet should be shoulder width apart
  - » Stand on firm ground
- Disadvantage
  - » Provides limited function
  - » Not effective in line construction
- Sharpening
  - » Sharpen back 2½ inches on each side with even bevel on both sides

Courtesy of Brian Henington

(continued)

Adz Hoe

- Best Uses/Characteristics
  - » Grubbing and trenching
  - » Fireline construction
  - » Great for cup trenches
- How to Use
  - » Similar to Pulaski
- Sharpening
  - » Bevel grub edge ⅜″ wide on a 45° angle. Inside edge only

© Brigida Soriano/Shutterstock.com

Scraping Tools

Scraping tools are effective in fireline construction and typically used to support cutting tools. They are also invaluable in mopup activities.

Fire Shovel

- Best Uses/Characteristics
  - » Most used fire tool across the United States
  - » Digging, scraping, smothering
  - » Throwing dirt
  - » Effective in breaking up berm created during fireline construction
  - » Effective in mopup activities
- How to Use
  - » Hold securely
  - » Feet should be shoulder width apart
  - » Scraping: use your knee as a pivoting point
  - » Stand on firm ground
- Sharpening
  - » Sharpen to the point, ½″ from heel. Sharpen entire cutting edge around shovel.

Courtesy of Brian Henington

(continued)

McLeod

- Best Uses/Characteristics
  - » Fireline construction in timber litter or grass
  - » Raking tines are very effective in certain fuel types.
- How to Use
  - » Hold one hand near the end of the tool
  - » Feet should be shoulder width apart
  - » Cut and pulling motion similar to a rake
- Disadvantage
  - » Not effective in rocky terrain
- Sharpening
  - » Maintain 45° cutting edge on flat scraping surface

Courtesy of Brian Henington

Combination Tool (Combi)

- Best Uses/Characteristics
  - » Digging, scraping
  - » Working between rocks
- How to Use
  - » Hold tool firmly
  - » Feet should be shoulder width apart
- Sharpening
  - » Stick the pick end in the ground
  - » Sharpen at a 45° angle

Courtesy of Brian Henington

Rhino Tool

- Best Uses/Characteristics
  - » Great for scraping or trenching
  - » Good in mopup activities
- How to Use
  - » Keep firm grip, feet shoulder width apart
  - » One hand toward the end of the tool
  - » Swing toward the ground to trench or use like a McLeod to scrap
- Sharpening
  - » Sharpen inside edge at 45° angle

Courtesy of Brian Henington

(continued)

Fire Rake

- Best Uses/Characteristics
  - » Raking debris
  - » Clean rocky areas
  - » Final tool in fireline construction
- How to Use
  - » Keep firm grip; feet shoulder width apart
  - » Pull rake toward you
  - » Kept weight off tines
- Disadvantage
  - » Only good for raking
  - » Tines can pick up embers

Courtesy of Brian Henington

Smothering Tools

Smothering tools are common in certain parts of the United States. The most common smothering tool is the swatter. Smothering tools can be effective in light fuel types.

Swatter (Flapper)

- Best Uses/Characteristics
  - » Grass fires or other light fuels
- How to Use
  - » Smoother the fire. Do not swing with hard vertical swatting. This will add oxygen to the fire
  - » Roll the swatter along the fire edge

Courtesy of Brian Henington

HAND TOOL ESSENTIALS

Regardless of the hand tool used on fires, there are established guidelines used to address safety concerns for the user. This section focuses on these essential elements. They include maintenance and inspection, carrying tools, working in the power zone, hand tool/pack lays, and the storage of tools.

Maintenance and Inspection of Tools

You should consistently evaluate your tool for any safety or performance issues. Hand tools are well built and should last for an extended period of time. However, firefighters should still evaluate their tool for any issues and repair minor damages. Try to repair damage; if you cannot, flag the tool, indicating it is not suitable for use. The list below provides a guideline for inspecting your hand tool.

Safety Note: Always wear gloves, safety glasses, and a long sleeved shirt when you are inspecting or working on your tool. This will eliminate cuts, metal shards, and/or splinters. In addition, the cutting edge of a tool should always be covered when not in use.

TABLE 12-1 *INSPECTION GUIDELINES*

Tool Head	Handle	Tool Rehab
Inspect Tool Head • Rust (weakens metal) • Damage to cutting head • Cracks, gouges • Check the wedge and ensure it is in place and tight • Cutting edge should be covered • Check any nuts or bolts • Lubricate moving parts (combi tool)	Inspect Handle • Smooth, no splinters (important to use gloves) • Secure to cutting head • Damage near cutting head • Condition of handle, especially cracks	• The cutting edge of a tool should always be covered when not in use • Sand handles • Treat handles if agency requires • Paint and mark tools for identification of ownership • Sharpen tool • Paint the tool head with a rust-prohibitive paint

Examples of Tools Not Safe For Use

Courtesy of Brian Henington

Damaged head with rust.

Courtesy of Brian Henington

Wedge missing. Extremely dangerous.

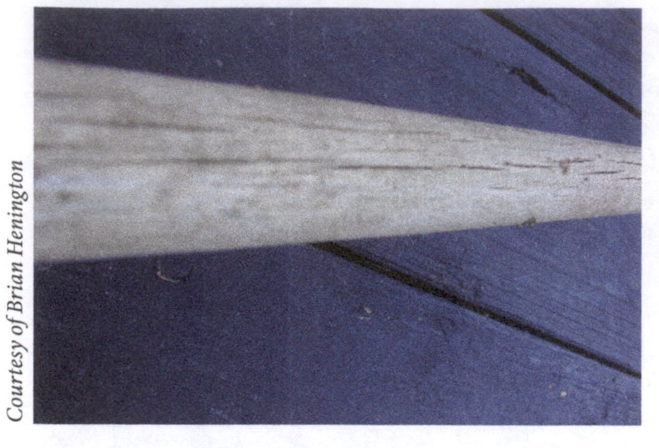

Courtesy of Brian Henington

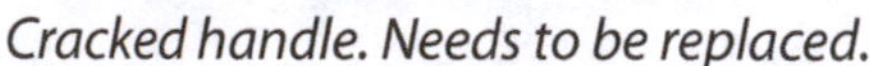

Cracked handle. Needs to be replaced.

Courtesy of Brian Henington

Flag damaged tools.

Carrying and Working with Tools

Hand tools are extremely sharp and some require an aggressive swing (e.g. Pulaski, felling ax, etc.). The firefighter needs to maintain a minimum of 10 feet when working next to other firefighters. The hand tool zone needs to be maintained at all times to ensure other firefighters are not injured by your activities.

Maintain a minimum of 10 feet between working firefighters.

The most important consideration when walking with tools is to cover the cutting edge to reduce the potential for cuts to the firefighter. After you have covered the cutting edge, the tool should be carried at the balance point (each tool will have a different balance point, but most are located close to the tool head). Two techniques for carrying hand tools are illustrated in the images below. Carrying a tool over your shoulder or resting it on your shoulder are not accepted techniques and should be avoided.

Courtesy of Brian Henington

By your side.

Courtesy of Brian Henington

Across your chest, in the power zone.

The firefighter should also consider the following safety elements when walking or working with a tool:

- Carry the cutting edge away from the body. Do not have the ax edge of a Pulaski pointed toward your body.
- On steep slopes, carry the tool on the downhill side.
- If you fall, throw the tool far enough away that you will not fall on it.
- Pass the handle end first if you have to transfer tools from one firefighter to another.
- If you have to walk through areas where hand tools are being used, call out a signal. (Coming Through! Clear a Hole! etc.) Do not walk through the area until firefighters have acknowledged it is safe to proceed.

Work in Your Power Zone

The power zone of the human body is between the shoulders and waist. To avoid injury because of improper use, work with the hand tool in your power zone. This may require that you bend at the waist when you are working with short-handled tools such as the Pulaski.

Courtesy of Brian Henington

Tool and Fire Pack Storage on the Fireline

There are several techniques to store your tools when not in use. The most preferred and safest technique is a tool lay. A tool lay requires that all similar tools be placed on top of each other. Most crews will have an organized approach to this concept. For example, as the crew breaks for lunch, the number one person on the crew will establish a tool lay. He will lay his Pulaski down. The second person in line will lay her Pulaski on top of the first Pulaski. The third person in line will lay her fire shovel next to the Pulaskis, but not on top of them. The fourth person will lay his shovel on top of number three's shovel. This process will continue until all hand tools are laid on the ground. As the crew returns to work, the last person in line will pick up his or her tool first. This will continue until the final tool is picked up. This technique addresses safety concerns of hand tools on the ground as well as maintaining accountability of the hand tools.

Courtesy of Brian Henington

Courtesy of Brian Henington

The second approved technique to store a tool when not in use (e.g. breaking for lunch) is to prop the tool against a tree or rock.

The key to both of these techniques is not to leave tools lying around where they can harm other firefighters.

The image above depicts a pack lay. The fireline packs are stored in a similar manner to a tool lay. Again, the reason for this is to provide organization and accountability of the equipment. Fireline packs can become a tripping hazard; thus the importance of this technique. The image below depicts a pack lay performed by the City of Santa Fe Fire Department's Atalaya Hand Crew.

Sharpening

Keep your tool as sharp as possible. This will reduce your physical exertion and improve the effectiveness of the tool. "A sharp, properly maintained tool is essential for accomplishing the job. A dull tool is dangerous and makes work more difficult."[1] It is also recommended that you review the manufacture's recommendations for sharpening the tool. The information below highlights essential sharpening considerations:

- Always wear gloves and a long sleeved shirt.
- Always wear safety glasses when sharpening your tool.
- Use a flat mile file with a hand guard.
- Slide the file over the cutting edge in a smooth, consistent fashion.
- Most hand tools will require a 45° angle to maintain the correct bevel.
- Visually inspect cutting edge; do not run your finger over the cutting edge to check for sharpness.
- If a tool cannot be repaired, flag the tool to indicate it is not safe for use.
- If the tool is being stored, ensure the cutting edge is covered.
- When walking with your tool, cover the cutting edge.
  - » Cutting edge protection examples (use any of the following to protect the cutting edge):
    - Duct tape
    - Filament tape
    - Old hose or rubber sheaths
    - Plastic sheaths

Courtesy of Brian Henington

Sharpening hand tool with a flat file.

Courtesy of Brian Henington

Cover the cutting edge. Duct tape is used in the photo above.

[1] National Wildfire Coordinating Group, *Firefighter Training, S-130* (Boise: National Wildfire Coordinating Group, 2003) 6.18.

FIRING DEVICES

Wildland fire managers use a wide array of firing devices to conduct burnout operations or ignite fuels. Some of these devices are complex and require a specialized skill set to operate. For the purpose of this book, we will introduce you to the two common firing devices: fusee and the drip torch. Experienced firefighters will typically use the firing devices identified below; however, it is very common for an entry-level firefighter to use these tools under close supervision of an advanced firefighter.

Courtesy of Brian Henington

Fusee

The fusee (or backfire torch) is an effective firing device and utilized throughout the United States. "A fusee is a hand-held disposable ground ignition device with a self-contained ignition system. Like common road flares, fusees generate a flame about 5 inches long"[2] All firefighters should carry fusees. Personally, I have witnessed the refusal of crew bosses to allow their crew members to carry fusees. Firefighters need to carry a fusee in case they have to burn fuels out in the preparation of a safe area, deployment site, and/or emergency safety zone.

Fusees are considered hazardous materials as they meet the requirements for combustible solids. They contain phosphorous, which can create severe burns to the firefighter. They should be stored in an appropriate manner and not stored next to flammable liquids such as gas or diesel.

TABLE 12-2 ***FUSEE ESSENTIALS***

Components	Special Features
• Safety Cap • Wrapper (waxed) • Handle (ferrule) • Striker (igniter)	• Ignition surface is protected • Burns for 10 to 18 minutes
Safety Considerations	
• Flame temperatures: 1400° F • Dripping sludge (wax) from burning fusee • If damaged or punctured, dispose of fusee • Ensure cap is in place at all times • Keep fusee away from heat • Store in cool, dry place	Courtesy of Brian Henington

[2] National Wildfire Coordinating Group *Interagency Ground Ignition Guide* (Boise: National Wildfire Coordinating Group, 2011), 71.

Fusee Use

Follow the steps below to properly ignite a fusee. **Note:** Your instructor should physically show you how to ignite and use a fusee.

Courtesy of Brian Henington

Step 1: Hold fusee by handle and ensure you have all the required PPE.

Courtesy of Brian Henington

Step 2: Remove striker cap.

Courtesy of Brian Henington

Step 3: Scratch striker against the ignition surface. Strike away from your body. Turn your head when striking.

Courtesy of Brian Henington

Step 4: Hold away from body.

Courtesy of Brian Henington

Step 5: Hold fusee toward the ground. (Avoid breathing smoke and vapors.)

Courtesy of Brian Henington

Step 6: Keep on the burn side (black) of the fireline. Do not stare at the bright flame.

Courtesy of Brian Henington

Step 7: Extinguish by striking against the ground.

Drip Torch

The drip torch is the preferred ignition device when burning out fuels. The National Wildfire Coordinating Group provides a detailed explanation of a drip torch:

> *A drip torch is a ground ignition device that has a fuel tank, a spout (also known as a burner), and an igniter with a wick. The drip torch tank is filled with a mixture of gasoline and diesel. The wick is ignited and drip torch fuel is poured out of the tank, through the spout, and past the burning wick. The burning wick ignites the drip torch fuel, which starts the fire.*[3]

The national standard for drip torches is to mix four parts diesel to one part gasoline. The standard tank size is 1¼ gallons.

[3] Ibid., 17.

TABLE 12-3 *DRIP TORCH ESSENTIALS*

Components	Special Features
• Fuel tank with handle • Breather valve • Spout and nozzle • Wick/wick holder • Discharge plug • Tank cover/gasket • O-ring (tank lock ring)	• Mixture: 4 parts diesel/1 part gas • Extinguish when not in use • Carry by handle only
Safety Considerations	
• Adding gas/diesel to the fire • Putting too much fire on the ground at any given time • Exceeding gas/diesel ratio • Reversing the fuel ratio can cause an explosion • Keep torch away from your body, fire clothing, and boots • Do not open or fill the torch next to active flames	*Courtesy of Brian Henington*

Drip Torch Use

Follow the steps below to properly use a drip torch. Note: Your instructor should physically show you how to safely use a drip torch.

Courtesy of Brian Henington

Step 1: Shake torch to mix fuel before use.

Courtesy of Brian Henington

Step 2: Unscrew the discharge plug and unscrew the lock ring.

Courtesy of Brian Henington

Step 3: Remove spout from fuel tank and allow fuel to drain back into the tank.

Courtesy of Brian Henington

Step 4: Place spout upright. Wick should face opposite direction of the handle.

Courtesy of Brian Henington

Step 5: Screw the lock ring back to the tank to secure spout. Fasten tightly.

Courtesy of Brian Henington

Step 6: Open breather valve.

Courtesy of Brian Henington

Step 7: Wipe up spilled fuel. Do not use your glove or fire shirt.

Courtesy of Brian Henington

Step 8: Light the drip torch by:

- *Pouring fuel on ground and lighting the fuel. Stick wick into burning fuel on the ground.*
- *Sticking wick into surface fire.*

Courtesy of Brian Henington

Step 9: Carry drip torch upright until ready to use. (Firefighter Erik Nelson)

Courtesy of Brian Henington

Step 10: Tilt torch downward when using. (Firefighter Gino Romero)

Courtesy of Brian Henington

Step 11: Extinguish: Set the drip torch on the ground in a vertical position. Allow the wick to burn out on its own. Make sure the drip torch is away from any active flame, but is in the burn side (black) of the fire.

Courtesy of Brian Henington

Wildland Firefighter Jarrod Duran using a flare launcher during burnout activities.

Other Ignition Devices

In addition to the drip torch and fusee, another commonly-used ignition device is the existing fire itself. Firefighters can do this by gently pulling burning debris (with their tool) such as pine needles into the area identified to burn. Firefighters can also use matches and/or lighters. Remember, the number one way to fight fire is with fire itself. Ensure you have the appropriate methods to ignite a fire, but NEVER light any fuels on fire without the authorization from your supervisor and with assistance from an advanced firefighter or experienced firefighter.

Firefighters can also use more complex firing devices. Devices can include flares, pin flares, pneumatic torches, very pistols, flare launchers, fusee launchers, terra torch, propane torches, ping pong launcher and/or helitorches. All of these devices are considered advanced and should not be handled by firefighters who are not certified or trained in their use.

SUMMARY

Your wildland fire hand tool is one of the most important pieces of equipment you will use on the fireline. Learn the techniques and safety features of your assigned tool and maintain it in the best manner possible. Practice using a tool at your house to strengthen your ability to use the tool effectively. Practice sharpening the tool. This will prepare you for the tasks you will perform on the fireline.

Firing devices can be dangerous and present additional hazards to firefighters. Learn the safety guidelines for the fusee and the drip torch. Never ignite a firing device without your supervisor's approval.

KNOWLEDGE ASSESSMENT

1. Identify the three categories of wildland fire hand tools.
2. True/False: Wildland firefighters should not inspect their hand tools to ensure they are effective and safe.
3. When working on tools, the wildland firefighter should wear ______________________, ______________________, and ______________________.
4. Firefighters should maintain ______________________ feet between each firefighter while working and walking with hand tools.
5. What is the most important consideration when walking with tools?
6. What are the two approved methods to carry a tool while you are walking?
7. A firefighter should hold the tool and work with the tool in the ______________________.
8. True/False: A sharp tool improves the effectiveness of a hand tool and reduces the wear and tear on the firefighter.
9. Why should every firefighter carry at least one fusee?
10. What is the fuel mixture for a drip torch?

BIBLIOGRAPHY

National Wildfire Coordinating Group. *Firefighter Training, S-130*. Boise: National Wildfire Coordinating Group, 2003.

National Wildfire Coordinating Group. *Interagency Ground Ignition Guide*. Boise: National Wildfire Coordinating Group, 2011.

CHAPTER 13

WATER USE AND DELIVERY METHODS

Learning Outcomes

- Identify and explain the common hose appliances and fittings used on wildland fires.
- Demonstrate the techniques used to unroll and roll fire hose.
- Described the two types of hose lays used on wildland fires.
- Explain the three nozzles patterns used on wildland fires.
- Express the importance of foam during wildland fire suppression efforts.

Courtesy of Brian Henington

Key Terms

Appliances
Backpack Pump
Fire Hose
Fittings
Floater Pump
Foam
Nozzle Patterns
Pump
Portable Pump
Progressive Hose Lay
Simple Hose Lay
Water Sources

OVERVIEW

Water greatly improves the suppression efforts of firefighters. However, water is not always available in large quantities in certain parts of the United States. Because of this, firefighters must be creative in the way they deliver water to a fire. Water is typically delivered by engines or water tenders or can be delivered by helicopters or other aircraft. When water is needed on the ground, firefighters will develop systems to effectively delivery the wet stuff to the hot stuff.

By itself, water will not totally eliminate the heat of burning objects. Firefighters with hand tools should always support water application techniques. Furthermore, conserving water is a major consideration when water is not readily available. Learning how to apply water in the right manner and form will conserve water and make it more effective in suppression efforts.

WATER USE APPLIANCES AND COMPONENTS

The connections that attach hoses are called **appliances** or **fittings**. These components are essential during the construction of hose lays and other water delivery activities. The information below introduces you to the common appliances or fittings used on wildland fires.

Pressure Relief Valve  *Courtesy of Brian Henington* Purpose: Relieves excess pump pressure to avoid damage to the pump.	**Adapter**  *Courtesy of Brian Henington* Purpose: To connect hose or appliances of different threads.
Reducer  *Courtesy of Brian Henington* Purpose: To reduce or decrease the size of a hose or appliance.	**Double Male Coupling**  *Courtesy of Brian Henington* Purpose: Connect two female ends together.

(continued)

Check and Bleeder Valve

Courtesy of Brian Henington

Purpose: Keeps water from flowing back into the pump when the pump is turned off or stops.

Double Female Coupling

Courtesy of Brian Henington

Purpose: Connect two male hoses together.

Custom Made Hose Ties

Courtesy of Brian Henington

Purpose: Maintains hose together after the hose is rolled.

Caps/Plugs

Courtesy of Brian Henington

Purpose: Protects the threads on the end of hoses from damage or prevents debris from entering sensitive areas.

Tee Valve

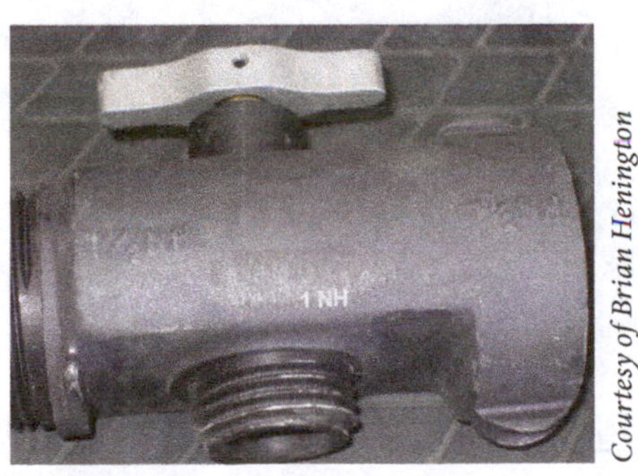

Courtesy of Brian Henington

Purpose: Allows a hose to be added to the main hose line. Common in progressive hose lays.

Gated Wye

Courtesy of Brian Henington

Purpose: Allows the main hose line to be split into additional lines.

(continued)

Shutoff Value (In Line)

Courtesy of Brian Henington

Purpose: Shuts off water.

Nozzles

Courtesy of Brian Henington

Purpose: Various sizes, types, and functions. The end of a hose lay.

Hose 1½″

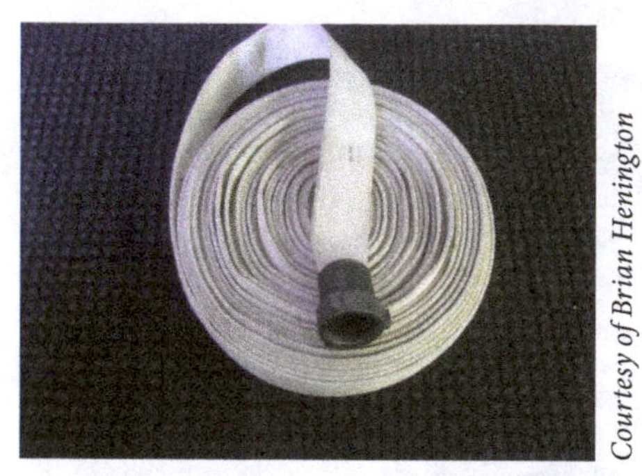

Courtesy of Brian Henington

Purpose: Typically used as main trunk line.

Hose 1″

Courtesy of Brian Henington

Purpose: Hose lays and booster lines.

Hose ¾″

Courtesy of Brian Henington

Purpose: Mopup operations; often called garden hose.

Gravity Sock

Courtesy of Brian Henington

Purpose: Captures water through gravity and delivers to the fire.

(continued)

Spanner Wrench

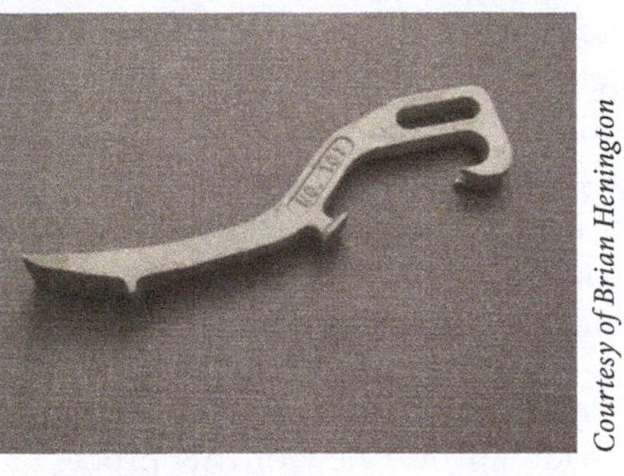

Courtesy of Brian Henington

Purpose: Loosen or tighten connections. Connect hoses and appliances by hand. Once snug, then tighten ¼ turn with spanner wrench. Avoid over tightening the connections.

Hose Clamp

Courtesy of Brian Henington

Purpose: Stops the flow of water in a hose lay.

Booster Hose

Courtesy of Brian Henington

Purpose: Usually on the reel of a fire engine.

Suction (Draft) Hose

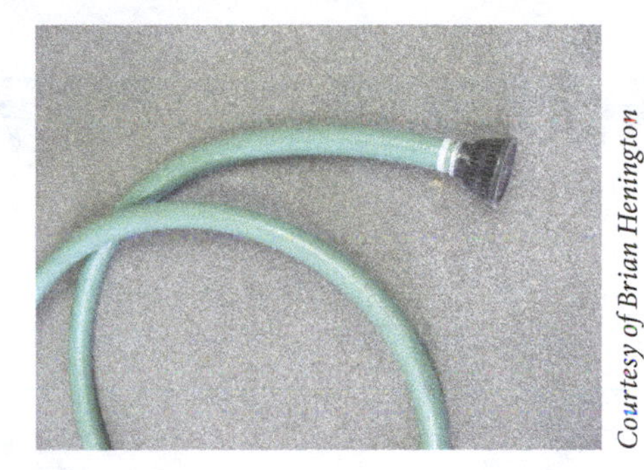

Courtesy of Brian Henington

Purpose: Hard line hose used to draft water from water sources.

Barrel Strainer (Foot Valve)

© Anton_Ivanov/Shutterstock.com

Purpose: Attaches to the end of the suction (draft) hose. Keeps dirt, mud, rocks, and other small material out of the pump.

Backpack Pump

Courtesy of Brian Henington

The backpack pump is typically used during mopup operations; however, it does provide a function during active suppression stages of a fire. The backpack pump holds five gallons of water and is contained in a nylon backpack with a plastic collapsible tank. A firefighter operates the system by engaging a trombone pump that is connected to the collapsible tank. The backpack pump is carried over the top of the fire pack, which adds an additional +/- 45 lbs. to the firefighter's gear.

The steps below are the steps used to effectively use a backpack pump. Your instructor should provide an overview of this process and you should demonstrate proper use to be eligible for national certification.

Feet should be shoulder width apart.

Courtesy of Brian Henington

Firefighter Manual Chavez.

Prime pump by moving the trombone back and forth (in and out).

Courtesy of Brian Henington

(continued)

Rotate the nozzle tip to adjust the nozzle pattern.

Courtesy of Brian Henington

Target the base of the flame or hot spot.

Courtesy of Brian Henington

Refill the collapsible tank with clean water. Dirty or muddy water will impact the effectiveness of the pump by clogging the nozzle.

The following safety measures should be used each time you use a backpack pump.

- Have other firefighters help you place the backpack pump onto your back.
- Make sure all the caps are tightly secured.
- Remove slack in backpack straps.
- Avoid stepping over large obstacles.

WILDLAND FIRE HOSE

The standard fire hose used on wildland fires is 1½", 1" or ¾" hose. The purpose of the smaller diameter hose is to conserve water when water is not readily available and utilize the effective pumping range of wildland fire pumps. The standard length of wildland fire hose is 50 or 100 feet. Individual hose lengths are often called sections or sticks. Hoses will have a male and female end. The male end is intended to connect to a female end of appliances/fittings or other hoses. As a rule, the male end of a hose should always be rolled toward the center or interior of a hose roll. The female end should connect to the male end of a pump or male end of other hoses.

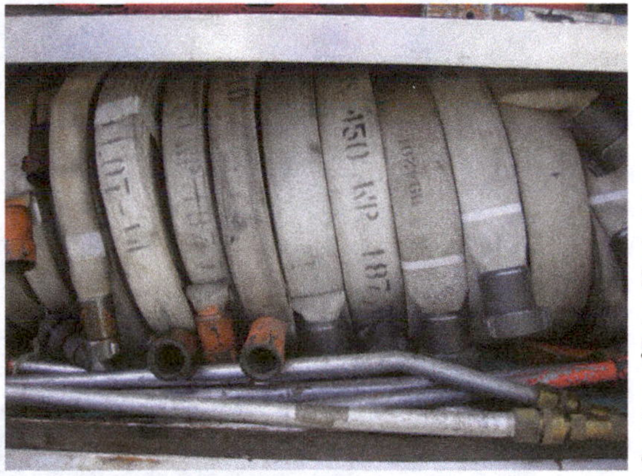

Courtesy of Brian Henington

Thread patterns must also be considered when working with additional fire engines or resources. The standard wildland fire hose thread types are: NH: National Hose; NPSH: National Pipe Straight Hose; or GHT: Garden Hose. Every engine should carry additional thread adaptors to ensure they are able to effectively work with other resources or in other geographical areas.

Hose should be maintained at high standards. As in the case of other fire equipment, fire hose is expensive. After use, fire hose should be drained and cleaned. Many fire agencies will have hose cleaning stations at their fire stations of offices. Allowing the hose to air dry will help reduce mildew. The ends of fire hoses should also be protected to avoid damage to the threads. The male end is more vulnerable than the female end; however, both are susceptible to damage.

Hose is also subject to damage during fire activities. Damaged hose should be flagged to identify product failure and should be replaced. Activities that are associated with damaged hose include burnt hose, hose cut by sharp rocks, vehicle damage, and/or hand tools hitting the hose.

Unrolling of Hose

There are several techniques used to unroll hose. Time may be of the essence, so some situations require fast deployment of the hose. The three types of hose unrolling are identified below.

Standard

The most common and basic technique is the standard unroll. This technique requires the firefighter to connect the female end to a pump or other hose and then walk out the male end. This technique is slower than the other techniques, but may be most appropriate in difficult terrain.

Courtesy of Brian Henington

Courtesy of Brian Henington

Underhand or Bowling

The second technique is to unroll the hose with an underhand roll (bowling ball). The firefighter should hold the female end in one hand or connect to a pump/hose and roll the hose like a bowling ball. The technique is ideal for larger diameter hoses.

Courtesy of Brian Henington

Courtesy of Brian Henington

Overhand Throw

The third technique is the overhand throw. The overhand throw requires the firefighter to hold onto the hose with one hand or connect to a pump. With the other hand, the firefighter holds the hose on his or her shoulder. The firefighter should identify a target on the ground about 15 feet in front. With a hard overhand throw, he or she throws the hose at the target. The hose should hit the ground and unroll to the desired destination. This technique is very fast but requires coordination and practice to use effectively.

Courtesy of Brian Henington

Courtesy of Brian Henington

Courtesy of Brian Henington

From a safety perspective, make sure all of your PPE is on (including eye protection and gloves). Shout out warnings that you are unrolling a hose (HOSE OUT! HOSE ROLLING! etc.). Finally, you want to make sure nothing of value (firefighters, structures, vehicles, etc.) is in the path you intend to use to unroll the hose. Unrolling hose can cause damage to threads so make sure the threads are protected before you unroll a hose. An entry-level firefighter is expected to be proficient in unrolling and rolling hose.

Rolling Hose

The ability to roll hose is a necessary skill for every firefighter. There are several techniques to roll hose. The common types are straight hose roll, watermelon, and/or figure 8. You should be able to demonstrate all three of these techniques before you become a certified wildland firefighter.

Straight Hose Roll

The straight hose roll is primarily used after a hose has been drained and cleaned after use on a fire. The hose is rolled in a tight roll and placed back onto the fire engine for the next fire assignment. **Note:** The male end should always be inside or to the interior of a hose roll. After the hose is rolled, secure hose with strapping tape, duct tape, parachute cord, or rubber bands.

Courtesy of Brian Henington

Courtesy of Brian Henington

Courtesy of Brian Henington

Watermelon or Melon Hose Roll

The second technique is the watermelon or melon roll. This technique is designed to remove hose from the field after use on a fire. The system allows the firefighter to tightly roll the hose in order to retrieve from the fireline. If the system is done correctly, a firefighter can carry four rolls of hose at one time. The photos below depict the stages for a watermelon roll.

Courtesy of Brian Henington

Step 1: Grab the male end of the hose in your right hand. Drain the hose. Start from uphill side to improve draining process.

Courtesy of Brian Henington

Step 2: Fold the hose over the top of the hose; approximately 20 inches.

Courtesy of Brian Henington

Step 3: Continue folding the hose on top of the 20 inches to make three overlapping layers.

Courtesy of Brian Henington

Step 4: Strengthen hose by turning the overlapping layers sideways.

Courtesy of Brian Henington

Step 5: Twist the hose six times over original folding. Pull out the slack.

Courtesy of Brian Henington

Step 6: Turn the hose lengthwise. Roll hose over and over, rotating slightly with each roll.

Courtesy of Brian Henington

Step 7: Pull hose toward you as you roll.

Courtesy of Brian Henington

Step 8: Hose should look like a football or watermelon.

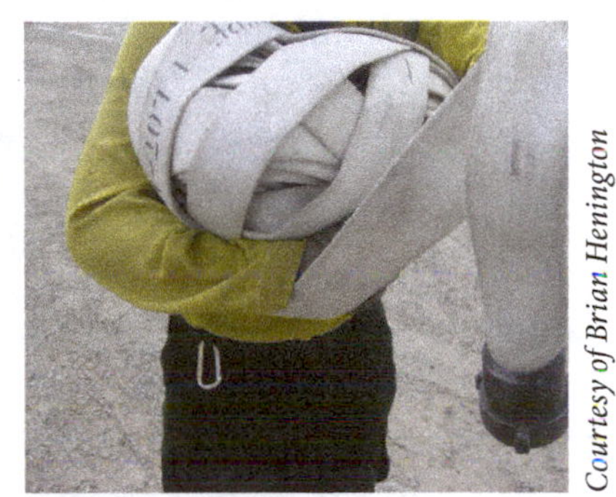

Courtesy of Brian Henington

Step 9: Stop about three to four feet from the end. Wrap the remaining three to four feet around the length of the hose. Pull slack out each time you wrap.

Courtesy of Brian Henington

Step 10: Continue wrapping the existing hose until your have about a 10" tail. Wrap the hose end through the existing hose to create a knot.

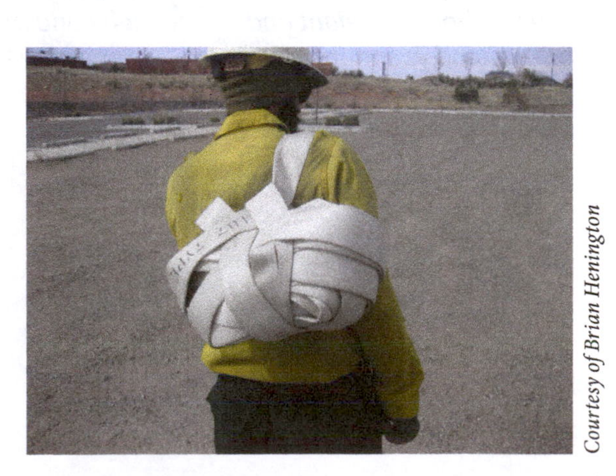

Courtesy of Brian Henington

Step 11: Once you are done, you can connect a watermelon roll to the first hose and sling over your shoulder.

Figure 8 Hose Roll

The third hose rolling technique is the Figure 8. This technique is the quickest hose rolling system and is often used when picking up hose from a hose lay and/or for reassignment to a new location. Again, the male end should be picked up first.

Courtesy of Brian Henington

Step 1: Grab the male end of the hose in your right hand. Hold your right palm toward the sky.

Courtesy of Brian Henington

Step 2: Take your left hand and run it down the hose. The back of your left hand should be upward.

Courtesy of Brian Henington

Step 3: Take your right hand and bring it down to your waist. Place the back of your right hand behind the remaining hose.

Courtesy of Brian Henington

Step 4: Lift your right hand back to your shoulder height and fully extend your right arm.

Courtesy of Brian Henington

Step 5: Take the back of your left hand and place it under the remaining hose. Lift to shoulder and fully extend your left arm.

Courtesy of Brian Henington

Step 6: Continue with your right hand, then your left hand, and so on. Continue until three to four feet of hose is remaining. Wrap this end of the hose side-ways around the hose you have just rolled.

Courtesy of Brian Henington

Step 7: Continue wrapping and tuck the female end through one of the loops to make a knot.

Courtesy of Brian Henington

Step 8: Connect the male end to the female end of the hose.

Courtesy of Brian Henington

Step 9: Place over your shoulder to carry hose.

PUMPING SOURCES

Wildland firefighters use three main pumping sources to deliver water to the fire. The main pumping source is the pump on a fire engine. These pumps vary in size and capability. They are designed to provide water from a mobile pumping platform and are used throughout the fire service as the most common pumping mechanism. The second pump type is a portable pump. Portable pumps will also have different capabilities and uses. The most common portable pump used on wildland fires is a Mark III pump. Firefighters will use these pumps to fill fire engines or deliver water to the fire. The third pump is the floater pump. These pumps are highly effective and can deliver a considerable amount of water to a fire or to fill water tenders or engines. The floater pump floats on top of the water and discharges water through a fire hose. Working with any of these pumps requires all PPE, but ear protection is especially vital as they are very loud and can cause long-term hearing damage.

Courtesy of Brian Henington

Engine Pump

Courtesy of Brian Henington

Portable Pump

Courtesy of Brian Henington

Floater Pump

WATER SOURCES

Firefighters may have to be creative in acquiring water for fire suppression activities. If you are lucky enough, you will work in areas with abundant water. The problem is that most areas do not have readily-available water. When this is the case, firefighters will transport water through water tenders or other mechanisms to the fireline. In addition, they will acquire the water from existing water sources. The list below shows a few examples of effective water sources. In certain parts of the country, you must get permission to use water from existing sources before use. We should be considerate and ask the land owner, farmer, or rancher if we can use their water. If we do use their water, we usually pay for it or replace it at a later time.

(continued)

Dry Hydrants

© kungverylucky/Shutterstock.com

Dry hydrants utilize water from ponds or lakes. Unlike normal fire hydrants, the system is not pressurized. During drought years, dry hydrants may not be functional because of the lack of water in a pond or lake.

Pumpkins

Courtesy of Brian Henington

Fold-A-Tanks

Courtesy of Brian Henington

HOSE LAYS

The National Wildfire Coordinating Group defines a hose lay as "any method or technique used to get water from the water source to the fire."[1] Water is delivered to a fire by two types of hose lays: simple hose lay and progressive hose lay. The information below provides a brief summary of the two types of hose lays.

Simple Hose Lay

A simple hose lay involves a pumping source, hose, and a nozzle operator. The simple hose lay is developed when additional hose lines are added between the nozzle and the pumping source. This hose lay does not have lateral extensions from the main hose line. The simple hose lay can involve the reel booster line or on a pre-connected attack line. The advantages of this hose line include:

- High pressure for the nozzle operator
- Ideal for mobile attack

[1] National Wildfire Coordinating Group, *Firefighter Training, S-130* (Boise: National Wildfire Coordinating Group, 2003), 8.14.

- Minimal hose is needed
- Charged line offers quick attack to the area of concern.

The disadvantage of this system is that it has to be clamped or shut off to add additional hose to the hose lay because it lacks gated wyes.

Progressive Hose Lay

The progressive hose lay is often referred to as the wildland hose lay. This hose lay can be explained as one pumping (or multiple) source, with hose and lateral hoses, and multiple nozzle operators. A progressive hose lay will have a main trunk line that has one or more branches or laterals from the truck line. An appliance/fitting is placed every 100 to 200 feet. This process allows multiple areas of target and use. The system is often used when mobile attack is not a viable option. The illustration below depicts a progressive hose lay.

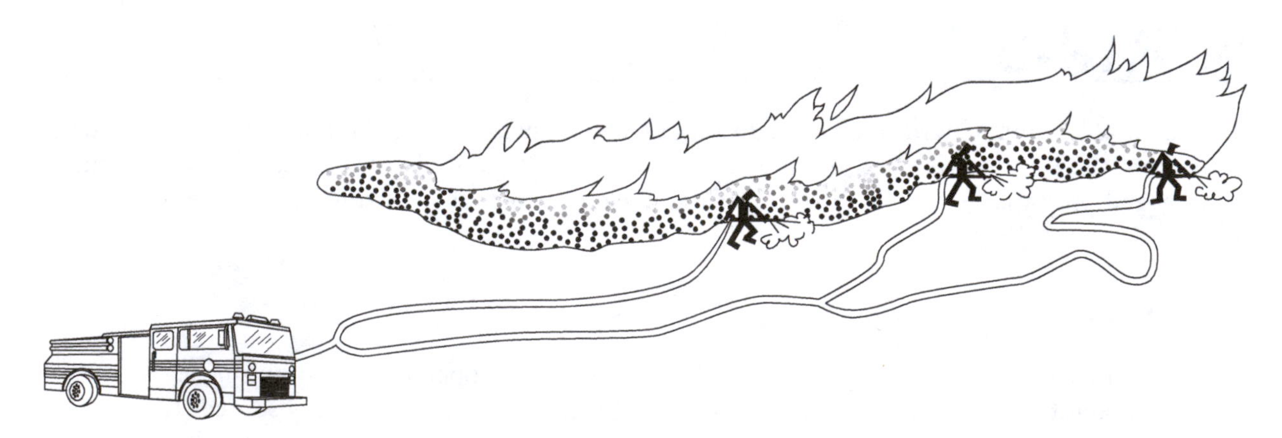

The benefits of a progressive hose lay are:

- Provides additional coverage to the target area(s).
- Provides for safety for each nozzle operator (additional nozzle operators can assist if conditions jeopardize the safety of other nozzle operators).
- Very effective in mopup situations because of the multiple coverage areas.

Negative aspects of a progressive hose lay:

- Long hose lays will require the appropriate pumping source to effectively deliver enough water pressure to each nozzle operator.
- This system requires large amounts of water. The system is only effective if fire managers have considered the amount of water needed for the hose lay to be effective.
- Long hose lays may require numerous resources, time, and energy to construct.
- Additional pumping sources may be required to ensure enough nozzle pressure is maintained for each nozzle operator.

NOZZLE PATTERNS

There are numerous types of nozzles available for wildland firefighters. The most common type of nozzle is the Forester Nozzle. This durable and tough nozzle is available in different diameter sizes. Other nozzles include combination barrel nozzles, gallonage nozzles, foam nozzles, ¾" mopup nozzles, and attack nozzles.

Courtesy of Brian Henington

Combination Nozzle

Courtesy of Brian Henington

Dual Gallonage Nozzles

Courtesy of Brian Henington

Forester Nozzle

A nozzle should provide three patterns. The first pattern is the straight stream. The straight stream is used to knock down fire intensity or for targets that are a distance away from the firefighters. Straight stream patterns do require an excessive amount of water and can fan the fire by providing oxygen to the fire. Straight stream patterns should be used for short durations and should be adjusted to other patterns to conserve water.

The second pattern type is the mist or fog pattern. This nozzle pattern is very effective in mopup situations as it saturates the target area and covers more surface area. The fog pattern uses less water and should be used when water availability is an issue. A fog pattern can also be used to provide a water barrier between firefighters and radiant heat.

The third pattern is the combination or spray pattern. This pattern incorporates both a straight stream and a fog/mist pattern. The pattern does not use as much water as a straight stream and covers more area than a straight stream. This pattern will require more water than a fog or mist pattern.

Courtesy of Brian Henington

Straight Stream

Courtesy of Brian Henington

Fog/Mist

Courtesy of Brian Henington

Combination

FOAM

Wildland firefighters utilize low expansion foams (Class A) to support fire operations. Foam can prolong the use of water and is highly recommended in firefighting efforts. It is highly effective as it improves the cooling effect of water and decreases water runoff. It will also adhere to the target area longer than water, which will increase fuel moistures. Foam will reduce the available oxygen to the burning material. Foam is also effective in structure protection efforts as firefighters can pretreat structures and leave the area when the fire intensity requires them to do so. Foam's effectiveness is further improved by the use of compressed air foam systems (CAFS) or air-aspirating nozzles.

© Dale A Stork/Shutterstock.com

Courtesy of Brian Henington

SUMMARY

The use of water during fire suppression efforts is highly effective. Delivering water to a fire may require additional skills and knowledge to ensure adequate supplies are maintained. It is important to understand the hose lays used on wildland fires and how to use the appropriate fixture or appliance when developing hose lays. In addition, every firefighter must understand how to properly roll, unroll, and maintain fire hose in the best condition available. You should constantly practice the techniques identified in this chapter to improve your effectiveness as a wildland firefighter.

KNOWLEDGE ASSESSMENT

1. How many gallons of water does a backpack pump hold? How much extra weight is this for the firefighter?
2. What are the four safety considerations when working with a backpack pump?
3. Individual hose lengths are referred to as ________________.
4. The three standard thread type for wildland fire hose are ________________, ________________, and ________________.
5. What are the three techniques used to unroll fire hose?
6. What are the three techniques used to roll fire hose?
7. What direction should the male end of a hose be facing when using a standard hose rolling technique?
8. What are the three main pumping sources used on wildland fires?
9. Identify the two hose lays used on wildland fires.
10. What are the three nozzle patterns used during suppression activities?

EXERCISES

1. Practice the three hose rolling techniques identified in this chapter.
2. Practice the three hose unrolling techniques identified in this chapter.

BIBLIOGRAPHY

National Wildfire Coordinating Group. *Firefighter Training, S-130.* Boise: National Wildfire Coordinating Group, 2003.

CHAPTER 14

THE SUPPRESSION OF FIRES

Learning Outcomes

- Explain the basic standards used when constructing fireline.
- Discuss the importance of fireline integrity.
- List the types of fireline construction used on wildland fires.
- State the two standard hand line construction methods used on wildland fires.
- Describe the blackline concept and how it relates to safe firefighting activities.
- Explain the importance of patrolling a fire and how these efforts support the overall suppression strategies used on wildland fires.

Courtesy of Brian Henington

Key Terms

Black Side
Blackline Concept
Bump-Up
Burning Out
Direct Attack
Fireline
Fireline Integrity
Flanking/Parallel Attack
Green Side
Indirect Attack
One Lick
Patrolling

OVERVIEW

Controlling a fire can be accomplished through several techniques and concepts, but the firefighter on the ground is the ultimate key to containing and extinguishing a fire. The activities that firefighters use to control a fire are called suppression actions. Suppression is basically the physical activities that firefighters perform to control a fire. The National Wildfire Coordinating Group explains suppression as "Management action to extinguish a fire or confine fire spread..."[1] This concept can be further explained as the actions necessary to extinguish a fire.

SUPPRESSION ELEMENTS

The basic method to control a fire is to remove one or more components of the fire triangle. As we explained earlier in this book, the element that firefighters can modify, manage, or mitigate the most is fuel. Oxygen is almost impossible to control on the larger scale, as it is more than abundant in the atmosphere. On a smaller or micro scale, we can control oxygen in the fire triangle by depriving it from burning fuels. This can be done with water, foam, dirt and/or fire gels. The second component of the fire triangle is heat. Heat is also hard to control on a 100°F day, but again, we can control it on a micro scale. Water, foam, dirt, and/or fire gels are some mechanisms we can use to impact the available heat for a fire.

Firefighters primarily strive to control the fuel component of the fire triangle. This is primarily accomplished by constructing fireline or using control lines. The concept of fireline specifically targets the fuel element of the fire triangle as it removes available fuel from the advancing fire. The next section of this chapter focuses on fireline importance and measures used to protect the integrity of the fire or control line.

FIRELINE CONCEPTS

Courtesy of Brian Henington

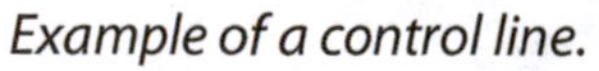

Example of a control line.

Courtesy of Brian Henington

The hand crew is constructing fireline.

[1] National Wildfire Coordinating Group, *Glossary of Wildland Fire Terminology* (Boise: National Wildfire Coordinating Group, 2014), 170.

A fireline is the primary mechanism used to suppress a wildland fire. A fireline occurs during the life of the fire. It is constructed by firefighters using hand tools, equipment, machines, explosives, and/or wetting or chemical agents. During active suppression activities, firefighters can also use a control line to inhibit the movement of a fire. The United States Forest Service defines a control line as "All built or natural fire barriers and treated fire edge used to control a fire".[2] Control lines can be natural, such as a river or overgrazed pasture, or can be man-made, like a road. As an entry-level firefighter, the two terms may be confusing, but the goal is to comprehend that some type of line must be in place before a fire is considered contained or controlled.

Black: Area inside the fireline
Green: Area outside the fireline

Another important concept regarding fireline terminology is the identification used by firefighters to reference the burnt and unburnt side of the fire or control line. The **green side** of a fire is considered the area outside of the control/fireline or the area that firefighters do not want to burn. The term used to describe the area inside the fireline/control line is called the **black side**.

Fireline Standards

Fireline standards can vary depending on a wide range of variables associated with the fire. Extreme fire behavior requires large firelines, whereas, small-scale, low intensity fires would not require a wide fireline. The fireline width and standards will be determined by a crew boss or other fire managers. Your role as a firefighter is to implement the identified standards and construct the fireline to the required protocol. In general, fireline should be 1 ½ times wide as the sustained flame lengths. As an example: A fire producing consistent two-foot flame lengths should have a three-foot constructed fireline. Fires that produce extensive flame lengths will require a wider fireline; however, there are other techniques used to reinforce or widen the line that do not require massive natural resource damage.

The information below should be used as a standard when constructing fireline.

Width: Fireline should be 1½ times wider than the consistent flame lengths.	 *Courtesy of Brian Henington*
Mineral Soil: Fireline should be dug to mineral soil. Mineral soil is soil that does not have any remaining biomaterials in it. These materials can support combustion and should be removed from the fireline.	 *Courtesy of Brian Henington*

(continued)

[2] United States Forest Service, *Fire Terminology*, accessed February 2, 2015, http://www.fs.fed.us/nwacfire/home/terminology.html#C.

Infinite: Fireline should be infinite. If the fireline is three feet wide on the ground, then it needs to be three feet wide at 40 foot tall. Branches from both sides of the fireline should not be touching. Chainsaw teams should remove these trees or branches.

Courtesy of Brian Henington

Fireline Integrity

The protection of a fireline is often called fireline integrity. The firefighters' objectives are to ensure their fireline can withstand an advancing fire or support burnout operations. The fireline should be clean of any burnable material and should not have ladder fuels adjacent to it. Heavy down and dead fuels and snags should also be removed to ensure the fireline is functional. The functionality of the fireline depends on how the line was constructed and the existing threats to the fireline.

Firefighters will reinforce the fireline and eliminate the threats to the control line. Furthermore, patrols can also locate threats and immediately mitigate them. The key is to ensure that the integrity of the fireline is in place and maintained as a priority. The information below depicts the threats to the fireline and concepts to mitigate those threats.

- Spot fires are an indication of extreme fire behavior. Convective lift of fire brand/embers or rolling debris on steep slopes are associated with spot fires. Holding and patrol efforts are used to identify spot fires. Once a spot fire occurs, firefighters should immediately suppress the fire and flag a location from the fireline to the spot fire to communicate its location. Crews should continually check the spot fire locations to ensure they are controlled and/or extinguished. Aircraft is also very useful in identifying spot fires for resources on the ground.
- Creeping is minimal fire behavior but can threaten the fireline because fire activity may not be obvious to the firefighter. Roots from snags or rotten trees are often associated with creeping fires crossing the fireline and igniting fuels outside the control line (green). Patrol and reinforcement efforts should be used throughout the life of the fire to identify any potential for this issue.
- Radiant heat is generated by fuels burning next to the fireline. This can preheat fuels on the green side of the fireline. Patrol efforts should be used to identify these concerns.
- Fireline construction on steep slopes can create several issues for firefighters. The main issue is rolling debris. Rolling debris can cross the fireline and ignite fuels on the green side of the fire. If fuels are ignited below the fireline, the expectation is that fire will spread fast uphill, which can eliminate available escape options for firefighters. A cup trench is required while constructing fireline on steep slopes. The cup trench is basically a ditch that is designed to capture rolling material or debris that can ignite fuels below the firefighter. The steeper the slope, the deeper the cup trench should be.

Types of Fireline

There are several types of construction methods used to construct fireline. The information on the following pages identify the fireline types used on wildland fires.

Hand Line: Hand line is constructed by firefighters on the ground with hand tools. The hand line is the most common fireline used in suppression activities.	 Courtesy of Brian Henington
Machine Line: Any fireline constructed by a machine. This could include a wide range of equipment. The primary machine line is a dozer line; however, it is not uncommon to see machine lines constructed by road graders, masticators, leaf blowers, or tractor plows. Fire managers are using more and more machines to assist with line construction because of their power, efficiency, and fast production rates.	 Courtesy of Brian Henington
Wet Line: Wet line involves the use of water and/or foam. Wet line can be delivered by fire engines, tactical water tenders, hose lays, helicopters and/or fixed wing aircraft (airplanes).	 Courtesy of Brian Henington
Retardant Line: Involves any man-made chemical used to control the spread of a fire. Retardant or fire gels can be delivered through several mechanisms, but the primary methods are by air resources (helicopters or air tankers).	 Courtesy of Brian Henington
Explosives Line: Fireline explosives are used in certain parts of the United States. This technique can provide fast fireline construction but requires firefighters with extensive training and qualifications.	

(continued)

Black Line: Typically implemented during ongoing suppression activities. A black line is highly effective in suppression efforts. Fighting fire with fire is a highly effective and much proven concept used in suppression activities. This concept is explained later in this chapter.	 Courtesy of James Bostin
Natural Control Line: May include a fuel break, ponds, streams, overgrazed pasture, roads, or cold fire edge. Utilizing existing features to control fire spread is highly effective in suppression efforts and helps reduce natural resource damage. Fire managers will often use natural control lines as ideal locations to contain or control the spread of a fire.	 Courtesy of Brian Henington

Hand Line Construction Techniques

As mentioned before, hand line construction is the most commonly constructed fireline. There are two hand line techniques used to construct fireline. They include one lick (progressive) and bump-up (leap-frog). These techniques are ideal when firefighters are working as a team to construct fireline. The concepts focus on teamwork and team members pulling their weight to accomplish the task. The selected technique will be based on the situation, circumstances, or the method preferred by the crew or engine boss. The methods are described below.

One Lick (Progressive)

One lick or progressive fireline is the most common hand line construction technique used on a fire. This technique involves each firefighter improving the efforts of the firefighter in front of him or her. In this process, the first crew member pioneers (strikes) the ground first and constructs close to two to three inches of fireline. He or she hits the ground and takes a step and hits the ground again. This continues again and again. The second firefighter improves the area the first firefighter hit and widens the line to three to six inches. The third firefighter improves the second firefighter's efforts and extends the width of the line to eight inches. This continues until the last person in line removes any remaining flammable material with a scraping tool and/or fire rake.

Courtesy of Brian Henington

Firefighters can construct fireline very fast and efficiently using this system. A few considerations or concerns with this type of construction method include:

- The width of the fireline should follow the standard width identified by the crew boss or fire manager.
- Crews have a tendency to build the fireline too wide using this technique. Everybody wants to contribute to the team effort, so he or she can improve what the firefighter in front has done. When this is done, the end result may be a six-to-eight-foot wide fireline when a three-foot wide fireline was all that was needed. This ultimately creates an issue with repairing (rehabbing) the fireline.
- A crew boss may often split crews into squads and have them work together from certain locations to ensure the most efficient fireline construction method is being used and to eliminate lines that are too wide.

1st Firefighter pioneers fireline.

2nd Firefighter improves the efforts of firefighter #1.

3rd Firefighter improves the efforts of firefighter #2.

4th Firefighter improves the efforts of firefighter #3.

All Courtesy of Brian Henington

Bump-Up (Leap-Frog)

The second technique used to construct fireline is called bump-up or leap-frog. In this concept, each firefighter is responsible for a specific piece of the fireline. The firefighter pioneers the line in his or her area and improves the line to the desired width. Everything from the first to last effort in a specific area is done by an individual firefighter. Once a firefighter is done, her or she will leap-frog the first crew member in line and begin to construct fireline again. This will continue with all 18 to 20 members of a fire crew.

ATTACK METHODS

Wildland firefighters utilize three primary attack methods to suppress a fire. The main consideration of the three concepts is the relation or exposure of the firefighter to the fire. The three attack methods are: 1) Direct Attack, 2) Indirect Attack, and 3) Flanking/Parallel Attack. The key to any of these attack methods is effective communications. According to Jerome Macdonald, the Deputy Incident Commander of the Southwest Area Incident Management Team and college Fire Science Instructor:

The key to safe and effective operations is communications. Every firefighter needs to follow the five communications responsibilities listed in the *Incident Response Pocket Guide* (IRPG):

- Brief others as needed
- Debrief your actions
- Communicate hazards to others
- Acknowledge messages
- Ask if you don't know

Direct attack is the preferred method to attack a fire. This attack method is often referred to as "Keeping One Foot in the Black." Direct attack is "constructing a fireline right on the fire perimeter."[3] This method is ideal because it allows the firefighter to immediately witness fire behavior. The standard or Golden Rule that determines direct attack is the flame lengths. If the flame lengths are sustained over four feet tall, then direct attack is not safe. The heat release rate is too intense for the firefighter to remain safe. The benefits and negatives of direct attack include:

Direct Attack: keeping one foot in the black

TABLE 14-1 *ADVANTAGES/DISADVANTAGES OF DIRECT ATTACK*

Advantages	Disadvantages
• Safest place to work • Minimal area burned • No additional area is intentionally burned • May reduce potential for crowning	• Can be hampered by smoke, heat, and flames • Control lines can be long and irregular • Doesn't take advantage of natural or existing barriers

The concept of attack from the black is considered a direct attack method. As you remember, the safest location for firefighters to attack a grass fire is from the black. The Texas Forest Service, an affiliation of Texas A&M University, studied and researched the effects of attempting a frontal assault on a fire and determined the most appropriate location to attack a grass fire is from the black. Attacking from the black provides the following tactical advantages:

1. Attacking the fire from your safety zone.
2. On wind-driven fires, the smoke will move with the direction of the fire spread. Locating resources in the black reduces smoke inhalation and the medical concerns associated with smoke.
3. Firefighters can see hazards such as large rocks, terrain depressions, or other terrain related hazards better than they can in the unburned fuel.
4. Enforces several Standard Fire Orders and Watch Out Situations as well as immediately enacting two of the four considerations of LCES (safety zones and escape routes).

© Dwight Lyman/Shutterstock.com

There may be situations when grass fires cannot be attacked from the black. If you are caught in one of these situations, you need to ensure your Rules of Engagement are in place and that you return to the black as soon as it is safe to do so.

The second method of attack is indirect attack. **Indirect attack** describes firefighters constructing the fireline a distance from the active fire. If flame lengths exceed four feet in height, then indirect attack may be the ideal attack method. The other consideration for indirect attack incorporates the use of existing control lines to control fire spread. It makes sense to utilize an existing road a quarter mile away from the fire rather than construct a new line that will require extensive labor and effort to construct and repair (rehab). Indirect attack can compromise several Watch Out Situations and the Standard Fire

[3] National Wildfire Coordinating Group, *Firefighting Training, S-130* (Boise: National Wildfire Coordinating Group, 2003), 9.6.

Orders. The main violation is the #11 Watch Out Situation: Unburned fuel between you and the fire. In order to mitigate this, firefighters should burn out available fuel between the fireline and the main fire.

The final attack concept is the **Flanking/Parallel Attack** method. Flaking/Parallel Attack is best described as "constructing a fireline by working along the flanks from an anchor point. May be either direct or indirect."[4] The determination of direct or indirect attack will be based on the activity and intensity of the fire. As in the case of indirect attack, any unburned fuel should be burned out by qualified firefighters.

SUPPRESSION TECHNIQUES

Firefighters utilize several proven techniques to suppress wildland fires. On most fires, firefighters will use a combination of these techniques to contain and control a fire. This chapter identifies the standard techniques used on wildland fires; however, there may be specific techniques used in your specific geographical area that are not covered below. Become familiar with the techniques and understand how to implement them.

Burning out is intentionally igniting fuels inside the fireline to eliminate unburned fuels. This technique mitigates the concern of unburned fuel between you and the main fire; however, this technique should be performed only after special consideration, preparation, and coordination. Burnout operations should always be supported by holding crews or holding activities. Remember, a firefighter is adding more ignition and combustion to the fire. This may present considerable issues if it is not done correctly or by qualified firefighters.

The **blackline concept** ensures firefighters have a solid edge of black directly against the fire or control line. This technique can involve intentionally igniting unburned fuels between the control/fireline and the main fire or allowing the fire to burn to the fireline. The blackline concept supports the integrity of the fireline and provides for safety of firefighters. The goal is to prevent the fire from making an aggressive run at the fireline. We want the fire to bump the fireline; not to hit it with all of its intensity and force.

All Courtesy of Brian Henington

The photos above illustrate an example of the blackline concept. If your fireline is 2 feet wide and you burn out an additional 2 feet of fuels, then the fireline is now 4 feet wide. If you then support this concept with proper ignition techniques, you can widen your fireline from 4 feet to 8 feet, then 8 feet to 25 feet. Twenty-five feet can then be burnt to 50 feet. This would make your original 2-foot wide fireline a total of 52 feet wide.

---

[4] Ibid., 9.6.

Hot Spotting: The technique of hot spotting is used to reduce fire spread or protect certain values at risk. This technique should not be used if the Rules of Engagement and LCES are not in place. The main purpose of hot spotting is to briefly reduce fire potential, intensity, and spread.	 *Courtesy of Brian Henington*
Prepping Line: Prepping or preparing the fireline is often needed prior to burnout operations. Prepping the fireline can include removing ladder fuels, low hanging limbs, and heat sources such as stumps and logs next to the fireline. Removing snags can also be involved in prepping the fireline. The removal of snags should be done only by certified firefighters.	 *Courtesy of Brian Henington*
Holding Line: Burnout operations should always be supported by holding crews. Holding crews ensure that burnout activities do not cause the fire to jump the fireline and burn on the green side of the fireline. Holding crews may be directly on the fireline or may use spot fire potential distance calculations to patrol areas in the green a distance from the fireline. If spotting distance is predicted to be ¼ mile, then crews will patrol 500 to 600 meters from the fireline to detect any spot fires.	 *Courtesy of Brian Henington*
Cold Trailing: This technique is used on cold portions of the fire or on areas of the fire with minimal fire activity. Cold trailing involves a complete effort in extinguishing any burning material that threatens the integrity of the fireline.	 *Courtesy of Brian Henington*
Scratch Line: Scratch line is used when resources are limited and/or to reduce the fire spread until additional resources can support the fireline construction activities. Scratch line does not follow the 1½ time rule used in standard fireline construction. The line may be six inches wide. The key is that a scratch line must be supported with additional resources and should meet the general fireline standard width required to control the fire.	 *Courtesy of Brian Henington*

(continued)

Fireproofing Fuels: Fireproofing fuels involves protecting the green portion of the fireline. Firefighters can treat fuels outside the fireline with water, foam, fire gels, retardant, or remove flammable fuels. This concept helps protect the integrity of the fireline and reduces the potential for fire establishing itself outside the fireline. In addition, this concept may include removing any threats that are immediately adjacent to the fireline.

© Peter Weber/Shutterstock.com

ADDITIONAL SUPPRESSION FUNDAMENTALS

A firefighter must be familiar with several other suppression fundamentals to ensure the most proficient and safe firefighting operations take place. This first consideration has to do with firefighter efforts to support machine fireline.

Machine Fireline Considerations

Heavy machinery is highly effective in constructing wide and large firelines. Even though it is used as the primary firelines on large fires, it does create some concerns that firefighters on the ground should address. The issues are identified below.

- Remove any flammable materials that have been left in the fireline.
- Break up machine piles. Heavy equipment activities may leave piles of flammable debris. If a flammable pile is located, firefighters should break up the available fuel. The ideal situation would involve heavy equipment breaking up the piles and spreading the debris inside the black of the control line.
- Fireproof areas of concern: Eliminate any ladder fuels, stumps, or snags that threaten the fireline.
- Mopup the fireline: To secure the machine line, mopup one to two chains from the fireline. This will help protect the integrity of the fireline.
- Patrol the control line: Identify any threats or issues through patrolling efforts.

Courtesy of Brian Henington

Debris left next to this dozer line will have to be removed.

Safety Concerns with Equipment/Resources Moving on the Fireline

A majority of fire suppression activities occur directly on the fireline. Firefighters should expect to encounter hazards, heavy machinery, vehicles, and fire engines along the fireline. The procedures below identify safety concerns when working adjacent to fire engines and/or heavy machinery.

Concerns with Engines and Firefighters on the Ground

1. As a rule of thumb, stay away from moving fire engines if you are not assigned to the fire engine.
2. Move off the fireline if an engine is driving on the fireline.
3. Charged hose lays are under high pressure and can fail. Be aware of charged hose lay locations and avoid working with hand tools right next to them.
4. Do not take a nap on the fireline.

Concerns with Heavy Equipment and Firefighters on the Ground

1. As a rule of thumb, stay far away from heavy equipment.
2. Move at least 100 feet off the fireline when heavy equipment is approaching or passing your location.
3. Avoid working below heavy equipment on steep slopes. A good rule of thumb is to allow the equipment to complete its work before beginning or continuing our activities.
4. Operators may not be able to see you and often are focused on the task at hand. Do not enter the danger area of heavy equipment until the equipment has been stopped and the operator or other personnel have advised you it is safe.
5. Do not take a nap on the fireline.

Retardant Safety Procedures

Retardant is very dangerous to firefighters on the ground. It is used consistently during suppression efforts to help slow the spread of the fire. Retardant is primarily delivered by air tankers or helicopters. The aircraft delivers the retardant at very high speeds and altitudes. The retardant (which weighs more than 10 lbs. a gallon) falls to the earth very fast (150 mph or 219 feet per second). In addition, retardant drops increase the terminal velocity and energy of the retardant when it hits the ground. You do not want to be under retardant drops! Firefighters who do not understand how dangerous retardant drops are may think it is heroic or macho to be under a drop. This is foolish thinking. Retardant drops have killed firefighters before and should be considered extremely hazardous to firefighters on the ground.

When retardant is intended for your area, you should ensure the following safety precautions are in place:

1. Make sure fire managers and the pilots know you are in the intended drop area.
2. All firefighters should move out of the area. Move at least 200 feet from the drop area. (Perpendicular to the planned retardant drop or the targeted drop area. If the plane is dropping due north, the firefighter should move east or west.)
3. Stay away from snags, widow makers, or other aerial hazards.
4. Do not move into the area until you have been cleared to do so. Additional air tankers may be following the first air tanker and continue dropping retardant in the area.

5. Retardant is slick when it is on the ground.
6. Firefighters are needed to support retardant or reinforce it. Once the drops are over, then move into the area to reinforce the retardant.

When retardant hits the ground, it can snap trees, knock down firefighters, or throw them against dangerous hazards such as rocks, trees, or hand tools. It can also dislodge rocks or other debris, and create flying projectiles. There is a reason why we wear yellow fire shirts. Pilots are not supposed to drop retardant if they see yellow-shirted firefighters in the intended target area. But there are times that unexpected drops may occur in the area you are working. If this is the case, follow the national safety standards provided by the National Wildfire Coordinating Group.

1. Lie down facing the oncoming aircraft, helmet on with chin strap, feet spread, goggles (or eye protection) on.
2. Hold hand tool at side; grab something solid such as a rock, tree, or shrub to break the force of a drop. If available or possible, get behind tractors, brush guard, fire engines, etc.
3. After the initial drop, move out of the area until you have received assurance that no more drops are to be made in that area.[5]

PATROLLING THE FIRELINE

Patrolling a fire is part of the overall suppression strategy used on wildland fires. The National Wildfire Coordinating Group defines patrolling as "To go back and forth vigilantly over a length of control line during and/or after construction to prevent breakovers, suppress spot fires, and extinguish overlooked hot spots."[6] Patrolling fires is done to ensure that the fire remains inside the fireline and does not cross the line. Patrolling is also essential in identifying spot fires, trees that have fallen across the fireline, or any rolling material that has crossed the line.

Courtesy of Brian Henington

If you are selected by your supervisor to conduct a patrol, you must first establish reliable communications. You should also use a mobile LCES system that constantly evaluates the Rules of Engagement as you move through an identified area. Patrolling efforts should include at least two firefighters in case of injury or other issues that may occur to a solo firefighter.

Patrolling efforts are intended to identify areas of concern. If you locate any concerns, they should immediately be relayed to your supervisor. You should also flag the concerns. Some issues may require immediate action from your patrol team. For example, if a spot fire has been identified, you should start suppression efforts (once Rules of Engagement are in place) immediately. Contact your supervisor and request assistance. Flag a location to the issue and make sure you advise supporting personnel of the color of flagging used.

---

[5] Ibid., 9.14.

[6] National Wildfire Coordinating Group, Glossary, 134.

Patrolling efforts are intended to locate:

Spot Fires: Pose threat to firefighters. Immediate attention should be given to spot fires.	© Ulysses_ua/Shutterstock.com
Weak Areas: Weak locations of the fireline are typically found at bends or corners in the fireline. In addition, midslope line may also be weak and require constant patrols to ensure the fireline is holding.	*Courtesy of Brian Henington*
Hot Spots: Identify areas retaining heat.	*Courtesy of Brian Henington*
Slopovers: Fire has crossed the fireline due to inadequate construction techniques.	*Courtesy of Brian Henington*
Reinforce Line: Remove any ladder fuels, downed logs, and/or stumps next to fireline. Remember, always take the material (if it can be moved safely) at least one chain into the black.	*Courtesy of Brian Henington*

(continued)

Fallen Snags: If a snag in the black falls across the fireline, the tree must be bucked up and all parts of the tree returned back into the black, at least one chain from the fireline.

Courtesy of Brian Henington

Patrolling Systems

There are several systems used to patrol fireline and/or for spot fires. The most common method involves at least two firefighters who work an area assigned by their supervisor. Their purpose is to identify any issues that may create additional problems to suppression efforts. The information below provides a brief example of an organized patrol system.

- Ensure communication measures are in place.
- Ensure LCES is mobile and is reevaluated as you move.
- Have a plan in place and maintain organization through the patrol efforts.
- Work in pairs.
- Select a visual indicator at the beginning of your patrol area and at the end of your patrol area.
- Work between the two indicators in a slow-moving fashion, looking for any threats or other issues.
- Work back and forth until your reach your destination. Advise supervisor and return using the same technique.
- Patrol efforts should be slow and careful. The firefighter is trying to locate and mitigate any issues; not move through the area very fast and carelessly.
- Remember, if you locate a spot fire, you need to immediately report your finding and the location of the spot fire to your supervisor. You should begin to attack the spot fire using the Rules of Engagement. Remember to flag the spot fire from its location back to the fireline.

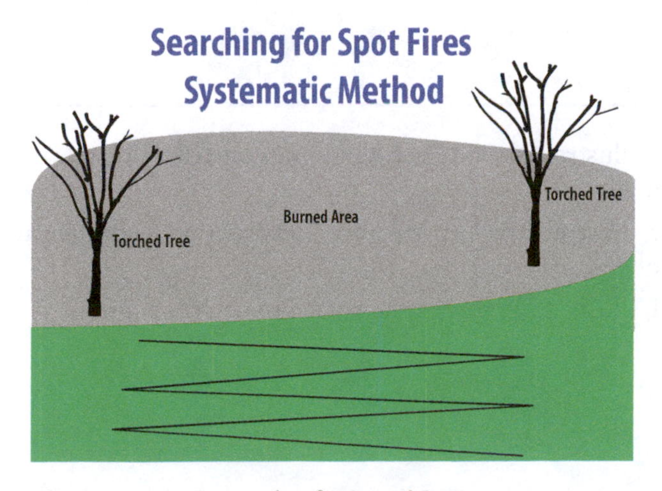

Figure 14-1 *Example of a Patrol System*

SUMMARY

Suppression efforts and techniques are essential to controlling a wildland fire. A firefighter may utilize several techniques to suppress a fire, but the underlying factor is the techniques must be safe. The integrity of the fireline should also be considered and efforts should be made to properly construct fireline and reinforce any issues or threats. Conducting patrol efforts to locate threats to the fireline and identify spot fires are also valuable and should be done on all fires regardless of size, complexity, or intensity.

KNOWLEDGE ASSESSMENT

1. What is the green side of a fireline and what is the black side of a fireline?
2. The basic principle for constructing fireline is: Construct fireline ________________________ times the sustained flame lengths.
3. Fireline should be dug or constructed to ________________________ soil.
4. What are the five types of fireline used to contain a wildland fire?
5. Describe the one lick hand line construction.
6. Describe the leap-frog hand line construction.
7. What are the three primary attack methods used on a wildland fire?
8. Firefighters cannot direct attack a fire if the flame lengths are over ____________ ft. tall.
9. What is the fire suppression concept in which fuels that remain between the main fire and the control line are burned out? This action provides for safety of firefighters.
10. What type of attack method is Attack from the Black?
11. What is the method of attack where fireline is constructed right on the fire perimeter? ("Keeping one foot in the black.")
12. Define the term "burning out."
13. Define the term "blackline concept."
14. Define the term "patrolling."

EXERCISES

1. On a piece of paper, illustrate the one lick hand line construction technique and list the pros and cons of this technique.
2. On a piece of paper, illustrate the leap-frog line construction technique and list the pros and cons of this technique.
3. On a piece of paper, illustrate a patrol system used to detect spot fires.

BIBLIOGRAPHY

Macdonald, Jerome (Deputy Incident Commander/Fire Science Instructor; Southwest Area Incident Management Team and Central New Mexico Community College Fire Science Program). Interview with author, March 23, 2019.

National Wildfire Coordinating Group. *Firefighter Training, S-130*. Boise: National Wildfire Coordinating Group, 2003.

National Wildfire Coordinating Group. *Glossary of Wildland Fire Terminology*. Boise: National Wildfire Coordinating Group, 2014.

United States Forest Service. *Fire Terminology*. http://www.fs.fed.us/nwacfire/home/terminology.html#C.

CHAPTER 15

WILDLAND FIRE SUPPORTING FUNDAMENTALS

Learning Outcomes

- Describe the importance of a hand-held, field-programmable radio.
- Recognize the significance of mopup activities and how they apply to the overall suppression strategy of a wildland fire.
- Express the dangerous associated with hazardous materials.
- Explain the importance of protecting the origin of a fire and the steps needed to observe and document essential evidence.
- Describe the significance of cultural resources and identify the steps used to protect these resources during fire suppression activities.

Courtesy of Mark Meyers

Key Terms

Cultural Resources
Dry Mopup
Field Programmable Radio Frequencies
Gridding
Hazardous Materials
Human-Caused Fires
Illicit labs
Natural-Caused Fires
Wet Mopup

OVERVIEW

A wildland firefighter is expected to comprehend and apply supporting fundamentals and concepts during fire suppression activities. These concepts include the use of a radio, extinguishment (mopup) of a fire, hazardous materials awareness, fire origin protection and the need to protect and preserve cultural resources. These fundamentals are part of the overall management of fires and should be considered part of the job duties and responsibilities of all wildland firefighters.

FIRELINE RADIOS

The main communication tool used on wildland fires is the **field-programmable, two-way (hand-held) radio**. There are several manufactures of hand-held radios, but the primary two-way radio used on fires is the Bendix King Radio. These field-programmable two-way radios are essential because they allow firefighters to input specific radio frequencies in order to establish communication with additional resources. In addition to hand-held radios, firefighters may also utilize radios in their assigned vehicle (mobile radio) or communicate with fire managers through radio dispatch centers. Establishing communications is essential to safe firefighting and a component of the Rules of Engagement and LCES.

Become familiar with the features of the Bendix King Radio illustrated in the image below.

Figure 15-1 *Field-Programmable Radio.*

Every fire agency is assigned a specific radio frequency or set of frequencies. **Frequencies** can be explained as similar radio waves. Radios transmit waves that other radios can receive if they are on the same frequency. Frequencies are assigned to fire agencies by the Federal Communications Commission and are intended for official use only.

When using a radio, firefighters should not cuss or use obscene language. You should also maintain prompt communications and professionalism at all times. Remember, everything said on a radio is heard and could be recorded. In addition, radio communication should always be conducted in clear text—no 10 Codes or CB talk. Finally, emergency traffic will have complete priority. An example of emergency traffic would be an injury to a firefighter. The initiator of the communication will announce to all personnel that emergency traffic will be broadcasted. All other resources are expected to clear the radio frequency and not use it until the original initiator has announced the emergency is over.

Effectively talking on the radio takes practice. It is not as easy as most people would assume. A firefighter's ability to effectively talk on a radio is a demonstration of professionalism, knowledge, and communication ability. Relaying information will require the proper use of jargon, techniques, and concepts. The firefighter should always remain calm and communicate with ease and confidence. Some guidelines for effective radio communication are identified below.

- Remain calm and talk clearly.
- Do not use any 10 Codes. Clear text only.
- Be courteous and respectful. Do not be disrespectful or rude.
- Respond to calling resources quickly and promptly.
- Talk normally. Do not yell or talk loud.
- Radio communication can be overwhelming at times. Write down messages or important information.
- Before you press the transmit button, think about you want to say before you say it.
- Be brief. Other firefighters are also using the radio.
- Use direct statements when communicating on a radio.
- Hold the microphone two to four inches in front of your mouth.
- Hold transmission button about one second before you speak and after you are done talking.
- Wait until other radio traffic is over before you begin talking.

Wildland firefighters use a standard process to transmit and receive messages. These steps are identified below:

1. When you are calling a resource, use the unit identifier or last name of the person you are calling and follow it with your unit identifier or last name. For example, if you are calling Engine 42 and you are Engine 61: "Engine 42 – Engine 61."
2. Engine 42 will respond with: "Go ahead Engine 61."
3. Relay the desired information and when you are done state: "Engine 61 clear." This tells other resources that you are done with your transmission and opens the frequency for use.

Although radios are an essential communication tool, they do present several challenges. The main issues with radios include lack of repeaters needed to handle communication, terrain obstructions, radio overload (too much radio traffic), and low or faulty batteries. Understanding these obstacles will allow the firefighter to implement solutions when any of these events are encountered.

MOPUP TECHNIQUES AND CONCEPTS

Mopup is an essential component of fire suppression and is often considered one of the last activities performed on a fire before a fire is considered controlled or extinguished. Mopup often lacks active fire behavior, but may possess the same risks and hazards that an active fire has. Firefighters have been injured or killed during mopup activities. Every firefighter should consider the mopup stage as dangerous and continue to prioritize the Rules of Engagement and LCES throughout the process. During this stage, fatigue and exhaustion play a major role in firefighters' ability to effectively evaluate their environment. Situational awareness must function during mopup as much as it does during the early and active stages of a fire.

TABLE 15-1 *COMMON HAZARD OCCURRENCES DURING MOPUP ACTIVITIES*

Hazard Occurrences		
• Snags	• Leaning trees	• Widow makers
• Stump holes	• Rolling material	• Burned roots
• Root holes/stringers	• Insects (gnats, bees)	• Hot ash, charcoal
• Water delivery activities	• Hose lays	• Steep terrain
• Rocks: exposure to prolonged heat may cause fragmentation when water is introduced		

WET MOPUP

Involves water and/or foam.

DRY MOPUP

Mopup without water and/or foam.

Mopup Techniques

There are several proven techniques used by wildland firefighters to effectively extinguish a fire. The techniques can involve water (wet mopup) or no water (dry mopup). Obviously, water improves the mopup process, but water is not always available to support mopup activities. The techniques for mopup are identified and described below.

Safety Note: Mopup activities have the potential to harm firefighters. Ensure that all PPE including eye protection is worn at all times.

Technique	Process	Visual Reference
Scraping	Removing burning material from unburned material. Example: the outer burning edge of a log should be scraped off and detached from the unburned portion of the log.	 © Alex_Po/Shutterstock.com
Digging	Exposes roots, stumps, and other burning material.	 Courtesy of Brian Henington
Mixing	Mixing unburned material with water/foam and/or dirt.	 Courtesy of Brian Henington
Separating	Separating burning materials from heat sources.	 Courtesy of Brian Henington

(continued)

Technique	Process	Visual Reference
Exposing and Spreading	Allowing relative humidity and cool temperatures to assist with mopup activities. Spread charcoal or duff.	 Courtesy of Brian Henington
Bone Piling	Create a bonfire. This will allow fire to consume heavy burning fuels and may expedite the consumption process.	 Courtesy of Brian Henington
Bucking	Bucking (cutting) large burning material into smaller material to facilitate the mopup process.	 Courtesy of Brian Henington
Felling	Felling hazard trees or snags by qualified and certified sawyers.	 Courtesy of Brian Henington
Turning	Turn material over to expose the burning underside of the log.	 © Chaikom/Shutterstock.com

(continued)

Water/Foam	Combine water and/or foam with dirt and the burning material. Highly effective. A mist nozzle pattern is the most effective. Always use hand tools to support water applications.	© Wyatt Rivard/Shutterstock.com
Mopup Kits	A long aluminum wand with a nozzle. Garden hose (3/4″) supplies water. Allows firefighters to penetrate burning material such as duff and timber litter. Always use hand tools to support water applications.	Courtesy of Brian Henington

Mopup Systems

Mopup protocol will be based on the needs and circumstances of the fire. Large fires will typically require activities to strengthen the control lines. Small fires may require complete mopup of all burning materials. Regardless of the size, mopup should first occur along the control line to maintain and prioritize the integrity of the fireline. Similar to patrol efforts identified in Chapter 14, mopup should follow an organized system or structure.

Courtesy of Brian Henington

The New Mexico Veterans Crew gridding from a control line.

The most common mopup process is called **gridding**. The National Wildfire Coordinating Group defines gridding as "To search for a small fire by systematically traveling over an area on parallel courses or gridlines."[1] Gridding should begin with an identified start and finish point. Crew members should work in pairs. The first pair will work along the control line and be responsible for 10 to 20 feet from the control line (from start to finish point). The second pair will work from 20 feet to 40 feet from the control line. The third pair would be from 40 to 60 feet. This would continue with the number of pairs needed to meet the mopup objectives.

This is just one example of a grid system. The point is that there needs to be an organized mopup system to ensure that all hot materials are detected and extinguished.

[1] National Wildfire Coordinating Group, *Glossary of Wildland Fire Terminology* (Boise: National Wildfire Coordinating Group, 2014), 93.

Identification Systems

Not all burning material is obvious. Firefighters can use their senses to help locate and identify burning materials or can use equipment designed to locate hot spots. Your sight is going to be the most useful for detecting hot materials. You should look for smoke, heat waves, white ash (area burned with considerable heat), steam, gnats, or stump/root holes. You can also smell burning materials. Some materials have a unique smell like cottonwood trees or rotten trees. Another important sense is touch. To detect heat, take off your glove and use the back of your hand to feel for heat. Of course, do not stick your hand into anything that has obvious heat.

Courtesy of Brian Henington

Technological devices can also assist in mopup, but they can be expensive and may not be used by some fire agencies. The devices are highly effective in locating hidden hot materials. Firefighters should use the equipment with care and protect it from damage. In addition, fire managers also use infrared imagery from aircraft to locate hot spots. Aircraft is highly effective and produces maps of intense heat or hot areas. Maps will be created that show these areas and fire managers can direct resources to deal with them. Drones with infrared imagery are used in some states to provide a similar function as aircraft. As drones become more affordable and effective, in the future fire managers may use drones to help firefighters determine areas of concern during mopup activities.

HAZARDOUS MATERIALS AWARENESS

Courtesy of Brian Henington

Firefighters should expect to encounter hazardous materials on a wildland fire. The International Fire Service Training Association defines hazardous materials as "any material that possesses an unreasonable risk to the safety of persons and/or the environment if it is not properly controlled during handling, storage, manufacture, processing, packaging, use, disposal, or transportation."[2] Hazardous materials or dangerous goods have the potential to severely harm, injure, or kill firefighters. If hazardous materials are detected, firefighters should immediately clear the area. They should notify the incident commander or appropriate overhead (using the chain of command). The appropriate fire manager will determine if hazardous material professionals should be called to handle the situation. You should avoid the area of concern because of

[2] International Fire Service Training Association. *Hazardous Materials for First Responders: 4th Edition* (Stillwater, Oklahoma: Oklahoma State University, Board of Regents, 2010), 9.

the great potential for harm and injury. The bottom line is: Do not engage hazardous materials if you are not certified, trained, or equipped. Leave the issue to the professionals who are trained, equipped, and certified to deal with hazardous material situations.

It is highly recommended that wildland firefighters take and be certified in an IFSAC (International Fire Service Accreditation Congress) course in hazardous materials awareness. The awareness level certification does not allow you to conduct offensive actions, but it increases your ability to detect hazardous materials and improves your overall situational awareness and hazard recognition ability.

Courtesy of Brian Henington

The most useful reference tool available for hazardous materials is the *Emergency Response Guidebook (ERG)*. The *ERG* is produced by the US Department of Transportation and provides essential information related to hazardous materials. The *ERG* is separated into several sections. The white pages of the *ERG* identify general information, placards, and trailer types. The yellow pages list the hazardous materials by numerical order of ID number(s). The blue pages provide an alphabetical order of the hazardous materials. The orange pages (considered the most important part of the *ERG*) identify the safety recommendations and general hazards. The green pages of the *ERG* contain information related to water reactive materials that produce toxic gases when water is introduced. Finally, the *ERG* will list important contact information for organizations that provide detailed information related to specific hazardous materials. Even though a majority of wildland firefighters are not certified or trained in hazardous materials awareness, you should still know how to use the *ERG* as a tool to help protect you from hazardous materials.

Hazardous materials can include a wide variety of substances or elements. The main categories of hazardous classes recognized by the US Department of Transportation are identified in the table below:

TABLE 15-2 ***HAZARDOUS MATERIALS CLASSES***

Class #	Class Description
1	Explosives
2	Gases (Vapors)
3	Flammable and Combustible Liquids
4	Flammable Solids, Spontaneous Combustibles, and Water Reactive
5	Oxidizers and Organic Peroxides
6	Toxic and Infectious Substances
7	Radioactive Materials
8	Corrosive Substances
9	Miscellaneous Hazardous Materials (Products, Substances, or Organisms)

As a rule of thumb, any material, container, or anything suspicious should be considered hazardous until it is determined by professionals that it is not a threat to firefighters. The steps below are a general guide to assist you if you encounter hazardous materials.

- GET OUT OF THE AREA!
- Move far enough away and stay upwind, upgrade, and upstream of the material.
- Contact the incident commander or designee (through the proper chain of command).
- Secure the scene to ensure other firefighters do not enter the area.
- If you can, try to identify the material by looking for placards, container labels, shipping papers and/or using the *ERG*. You may have to use binoculars to do this safely.
- Do not enter the area until authorized personnel have evaluated the scene and declared it to be safe.
- If a hazardous material scene is declared, allow the professionals to manage the incident and continue firefighting efforts to suppress the wildland fire. This will typically be handled as an "Incident within an Incident" or a specialized group will be created within the operations section of the incident to deal with the hazardous materials.

Another concern to wildland firefighters that involves a combination of criminal activity and hazardous materials is **illicit labs**. These labs may include drugs, chemical agents, explosives, and/or biological labs. Illicit labs are very harmful and can be located in remote areas, where wildland fires occur. Illicit labs may have extensive bobby traps or hazardous materials that can severely harm or kill firefighters. If you encounter anything strange, out-of-place, or suspicious, quickly retreat from the area (using the same route you entered) and contact the incident commander (through the proper chain of command). Secure the scene and deny entry until the site has been evaluated and deemed safe.

FIRE INVESTIGATIONS

Wildland fires are primarily started by two ignition types: **natural** and **human-caused**. The ignition source of a fire needs to be determined and documented. Fire managers track ignition data and use it for planning and prevention activities. In addition, human-caused fires can trigger criminal or civil prosecution. Fire misuse (accidents) and intentional (arson) fire starts will activate investigations that will be conducted by qualified and certified fire investigators. The National Interagency Fire Center (NIFC) is the primary agency that records and tracks fire ignitions for wildland fires. In a recent study of fire ignitions, NIFC has determined that an average of 62,631 wildland fires a year are caused by human activities. They have also determined that on a yearly basis, an average of 10,000 ignitions are caused by natural ignitions. NIFC has based this data on a 13 year average ranging from the years 2001 to 2013.[3]

The information below depicts some examples of ignition sources, including both human-caused fires and natural-caused fires.

[3] National Interagency Fire Center, *Lightning-Caused Fires and Acres (2001–2013) & Human-Caused Fires and Acres (2001–2013)* (Boise: National Interagency Fire Center, 2013), accessed March 20, 2015, http://www.nifc.gov/fireInfo/fireInfo_statistics.html.

Human-Caused Ignition Sources

Matches/Lighters © Phil McDonald/Shutterstock.com	Fireworks © dsmsoft/Shutterstock.com	Cigarette/Cigar Butts © PHOTO FUN/Shutterstock.com
Damaged or Downed Power Lines © Glynnis Jones/Shutterstock.com	Road Flares/Fusees Courtesy of Brian Henington	Vehicle Wrecks © Four Oaks/Shutterstock.com
Oil/Gas Flaring © John Kasawa/Shutterstock.com	Camp Fires © Matushchak Anton/Shutterstock.com	Debris Burning © ZanozaRu/Shutterstock.com

(continued)

Human-Caused Ignition Sources

Courtesy of Brian Henington

Equipment

© Andrea Slatter/Shutterstock.com

Welding

©Sergiy1975/Shutterstock.com

Firearms/Tracer Rounds

Natural-Caused Ignition Sources

© J.M.P.M Seijger/Shutterstock.com

Lightning

© Detchart Sukchit/Shutterstock.com

Spontaneous Combustion (Wood chip piles, hay bales, manure piles, etc.)

The duties of initial attack resources focus on three important concepts as they relate to fire investigations: 1) observe suspicious activities, 2) document such activities, and 3) protect the point of origin. If you notice, there is no mention of interrogating witnesses or suspects or determining what caused the fire. None of these activities are our responsibility. You are responsible ONLY for gathering and documenting information and turning it over to qualified fire investigators. In addition, if you are not a certified fire investigator, then do not contaminate the point of origin or disturb evidence. Our responsibility is to secure the point of origin and allow the fire investigators to do their job.

The information below describes the considerations firefighters should take in gathering and recording information to assist in investigation efforts and the steps necessary to protect a point of origin.

TABLE 15-3 *FIRE INVESTIGATION CONSIDERATIONS*

Observe	Document	Protect Origin
• Suspicious vehicles, ATV/UTVs, or people. • Children in area or leaving the fire area. • Vehicles driving fast or with lights off as they leave the fire area. • Downed or damaged power lines. • Human activity indicators: footprints, tire impressions, etc. • Fire behavior: color of smoke, fuels, terrain, etc. • Foreign objects such as fireworks, campfires, matches, etc. • Any other suspicious activity.	• Assumed point of origin on a map. Take GPS readings. • Photograph assumed point of origin and surrounding areas. • Photograph suspicious activity and/or foreign objects. • Record suspicious vehicles: color, make, model, license plate number, and any special features (bumper guard, headache rack, etc.) • Record suspicious people: gender, height/weight, ethnicity, hair color, eye color, clothing, etc. • Record fire behavior. • Note weather readings such as wind speed/direction, RH, temperature, etc. when you arrive. • Document on ICS Form 214- Unit Log and draw a map of the assumed point of origin and the surrounding area.	• Flag area. Flag perimeter around assumed origin. • Post guards if needed. • Eliminate fire activities such as hose lays, line construction, machine line etc. in the point of origin. • Do not touch or remove evidence. • If you have to enter the point of origin, enter at one point and leave at the same point. Identify the area where you entered for the fire investigators.

CULTURAL RESOURCES

Wildland firefighters conduct fire activities in areas of significant historical value that are referred to as cultural resources. A firefighter should make every effort to avoid activity that can damage these priceless resources. The National Park Service defines cultural resources as "… physical evidence or place of past human activity: site, object, landscape, structure; or a site, structure, landscape, object or natural feature of significance to a group of people traditionally associated with it."[4] In laymen terms, cultural resources are a link to the past and provide valuable information on cultures, religions, traditions, and activities. Cultural resources are classified into several categories. These categories are briefly described in the table below.

[4] National Park Service, *Cultural Resources* (United States Department of the Interior), accessed March 19, 2015, http://www.nps.gov/acad/learn/management/rm_culturalresources.htm.

TABLE 15-4 *CULTURAL RESOURCE CATEGORIES*

 © marekuliasz/Shutterstock.com Artifact: A portable material object made or modified by humans.[5]	 © Allison Herreid/Shutterstock.com Feature: An artifact that is not portable such as a rock structure or petroglyphs.
 © Andrea Visconti/Shutterstock.com Site: An area utilized or occupied by humans as indicated by the presence of cultural resources.[6]	 Courtesy of Brian Henington Ethnographic resources: Sites, structures, landscapes, objects or natural features of significance to a traditionally associated group of people.[7]

Cultural resources are protected by several laws. If a cultural resource is damaged, it may be lost forever. The key concept firefighters must remember is not to remove, take, touch, or damage any suspected cultural resource. Federal laws and state laws have severe penalties for damaging or stealing cultural resources. The main two federal laws designed to protect cultural resources are the National Historic Preservation Act of 1966 and the Archaeological Resources Protection Act of 1979. In addition to these two federal laws, certain states have very powerful laws or statues specifically designed to protect resources of cultural significance. As a firefighter, you should research and familiarize yourself with these laws to ensure that our fire activities do not violate these rules.

Cultural resources are of such value that fire managers use the expertise of cultural resource specialists (who are also certified firefighters) to identify potential cultural resources and provide mitigation

[5] National Wildfire Coordinating Group, *Firefighter Training, S-130* (Boise: National Wildfire Coordinating Group, 2003), 18.2.

[6] Ibid., 18.3

[7] National Park Service, *Cultural Resources.*

measures to protect the resource. A good rule of thumb is to avoid using suppression activities at the locations. The information below depicts certain fire-related activities that can create damage to cultural resources.

- Heavy equipment: Blades or tracks from heavy equipment can cause significant damage.
- Vehicles: Engines, water tenders, support vehicles, ATV/UTVs can create damage to cultural resources.
- Water delivery systems: Hose lays and high-pressure nozzle application can cause damage.
- Fireline construction: Can destroy or displace cultural resources.
- Aerial application: Retardant or water bucket drops can have negative impact due to the terminal velocity of the retardant and/or water.
- Staging areas or parking areas: Can damage resources by vehicle or human activities.

At some point in your career, you should expect to encounter sites, artifacts, or features. Some measures you can use to ensure cultural resources are maintained in their current condition are:

- Do not disturb the site.
- Take GPS points and mark the location on a map.
- Flagging can be used to identify the sites.
- Advise your supervisor who should advise fire managers.
- Avoid any tactical activities through or within the site.
- When in doubt, allow an expert (cultural resource specialist) to make the mitigation decision.

Overall, firefighters should make every effort to avoid damage to cultural resources and should understand the significant value these resources have to certain cultures, religions, and our society.

SUMMARY

Supporting fundamentals and concepts are an important component of wildland fire management. As an entry-level firefighter, you should familiarize yourself with the concepts covered in this chapter. Become familiar with proper radio use and mopup techniques used to control a fire. Also understand how dangerous hazardous materials can be to firefighters. Realize that the best course of action is to avoid situations that have hazardous materials. In addition, fire misuse and arson will trigger investigations. Do not contaminate or destroy any evidence. Learn the techniques to protect the point of origin of a fire. Finally, remember that cultural resources have great historical and cultural value. Understand the techniques and concepts to protect and avoid these priceless resources.

KNOWLEDGE ASSESSMENT

1. Wildland firefighters use what type of hand-held radio? What is the most common manufacturer of wildland fire radios?
2. What are frequencies?
3. Are 10 Codes allowed on a wildland fire?
4. Why is mopup important to fire suppression?
5. What are the two types of mopup performed on a wildland fire? Explain both.
6. What is the primary mopup system used on wildland fires?
7. True or False: A wildland firefighter should not expect to encounter hazardous materials on a wildland fire.
8. What is the bottom line on dealing with hazardous materials on a wildland fire?
9. What is the most useful reference tool when dealing with hazardous materials?
10. What are illicit labs? Why are they harmful to firefighters?
11. What are the two categories of ignition types?
12. What are the three primary duties related to fire investigations for initial attack resources?
13. Define a cultural resource.
14. List the measures you can use to ensure cultural resources are maintained.

BIBLIOGRAPHY

International Fire Service Training Association. *Hazardous Materials for First Responders: 4th Edition.* Stillwater, Oklahoma: Oklahoma State University, Board of Regents, 2010.

National Park Service. *Cultural Resources.* United States Department of the Interior. http://www.nps.gov/acad/learn/management/rm_culturalresources.htm.

National Interagency Fire Center. *Lightning-Caused Fires and Acres (2001–2013) & Human-Caused Fires and Acres (2001–2013).* Boise: National Interagency Fire Center, 2013. http://www.nifc.gov/fireInfo/fireInfo_statistics.html.

National Wildfire Coordinating Group. *Firefighter Training, S-130.* Boise: National Wildfire Coordinating Group, 2003.

National Wildfire Coordinating Group. *Glossary of Wildland Fire Terminology.* Boise: National Wildfire Coordinating Group, 2014.

CHAPTER 16

WILDLAND-URBAN INTERFACE CONCERNS AND CONCEPTS

Learning Outcomes

- Explain the issues associated with fighting fires in the wildland-urban interface.
- Describe the unique hazards that are associated with the wildland-urban interface.
- Identify the three tactical plans used during wildland-urban interface fires.
- List the structure triage categories.
- Restate the Wildland-Urban Interface Watch Out Situations.

Courtesy of Brian Henington

Key Terms

Combined Mode
Defensive Mode
Illicit Labs
Offensive Mode
Structural Triage
Wildland-Urban Interface

OVERVIEW

The suppression of wildland fires is complex and dynamic. Weather, fuels, and terrain can influence fire behavior that can directly impact firefighter safety. Complexity increases as fire behavior elevates. Wildland fire suppression can be complex by itself. Now add homes and civilians into the picture and complexity can skyrocket.

Welcome to the wildland-urban interface!

The wildland-urban interface can be one of the most dangerous fire suppression activities a wildland firefighter ever deals with. Understand that the safety of firefighters is the top priority. Second is the protection of civilians. Structures will be protected only if it can be done safely.

THE WILDLAND-URBAN INTERFACE (WUI)

The National Wildfire Coordinating Group defines the wildland-urban interface (known as WUI) as "The line, area, or zone where structures and other human development meet or intermingle with undeveloped wildland or vegetative fuels."[1] Firefighters are often tasked with protecting structures. Structures can include any value or improvement constructed or utilized by humans. Structures can be homes, apartment complexes, commercial buildings, out buildings, barns, trailer homes, etc. Firefighters will not base their protection measures on the structure's value, but will engage protection measures based on structural triage that focuses on firefighter safety and the structure's ability to sustain combustion.

The main concern with these fires is the elevated risk and hazards to firefighters. Firefighters need to remember that not every structure or home can be saved. According the *Fireline Handbook*, "Structures exposed to wildland fire in the urban interface can and should be considered as another fuel type."[2] The above statement is powerful and the public may not like it, but the point is that firefighters should not risk their lives for a structure. "Firefighters tend to place themselves at greater risk when battling wildland-urban interface fires in an effort to save homes."[3] Structures can be rebuilt; you cannot.

Matthew Pena, a college Fire Science Instructor and wildland firefighter, explains the importance of safety as it relates to interface fires, "Through all this we must still remain safe and remember to provide for safety first. This can become difficult to do at times but it is essential."

More and more fires are threatening homes. Today's wildland firefighters not only deal with remote fires, they deal on a regular basis with fires that threaten structures. Because of this, wildland firefighters may face difficult decisions regarding unfamiliar tactics and hazards. There is also the common philosophy among firefighters that all structures must be protected. This attitude is not acceptable. Firefighters have been injured or died trying to protect homes.

[1] National Wildfire Coordinating Group, *Glossary of Wildland Fire Terminology* (Boise: National Wildfire Coordinating Group, 2014), 187.

[2] National Wildfire Coordinating Group, *Fireline Handbook* (Boise: National Wildfire Coordinating Group, 2004), 10.

[3] National Wildfire Coordinating Group, *Firefighter Training, S-130* (Boise: National Wildfire Coordinating Group, 2003), 14.3.

Utilize our Rules of Engagement, including LCES, before any tactical action is considered. According to the *Incident Response Pocket Guide*, tactical actions in interface fires should be based on the following:

- Base all actions on current and expected fire behavior. Do this first!
- An estimate must be made of the approaching fire intensity in order to determine if there is an adequate safety zone and time available before the fire arrives.
- Due to the dynamic nature of fire behavior, intensity estimates are difficult to make with absolute certainty. It is imperative that firefighters consider the worst case and build contingency actions into their plan to compensate for the unexpected.[4]

UNIQUE HAZARDS

The urban interface has unique and very dangerous hazards that a typical wildland fire does not have. The hazards have the potential to cause great harm or death to firefighters. Hazards should be identified and mitigated prior to engagement by firefighters. If the hazards cannot be adequately mitigated, than avoidance may be the best solution. If you encounter an area of extreme and unique hazards, flag-off the access to the road to ensure firefighters do not enter the area.

Power Lines

Electricity is extremely dangerous to firefighters. In the history of our country, convicted criminals were sentenced to death by the electric chair. The electricity used during this process was 2,000 kilowatts. A single-phase power line (one power line on the top of the utility pole) is typically 7,000+ kilowatts. You should always evaluate a structure for power lines before implementing tactical activities. Energized electrical lines on the ground can kill firefighters.

The *Incident Response Pocket Guide* provides a clear set of guidelines for dealing with power line safety. You should become familiar with the safety precautions so you can avoid injury to yourself. The information below is taken from the *Incident Response Pocket Guide* to improve your knowledge on safety concerns as it relates to downed power lines.

- Communicate: Notify all responders of downed electrical lines. Obtain radio checkback.
- Identify: Determine entire extent of hazard by visually tracking all lines, two poles in each direction, from the downed wire.
- Isolate: Flag area around down wire hazards; post guards.
- Deny entry: Delay firefighting actions until hazard identification and flagging is complete and/or confine actions to safe areas.
- Downed line on vehicle: Stay in vehicle until the power company arrives. If vehicle is on fire, jump out with both feet together. Do not touch the vehicle. Keep feet together and shuffle or hop away.
- Always treat downed wires as energized!

Figure 16-1 *Safety Concerns with Downed Power lines.*

[4] National Wildfire Coordinating Group, *Incident Response Pocket Guide* (Boise: National Wildfire Coordinating Group, 2014), 12.

In addition to the above safety precautions, there are several other issues that firefighters should consider when working around power lines. They are identified in the table below.

TABLE 16-1 *CONSIDERATIONS WHEN WORKING AROUND POWER LINES*

• Heavy equipment should not work directly under power lines. • Avoid fueling vehicles, chainsaws, or filling drip torches under power lines. • Avoid direct water application to power lines. • Dense smoke can act as a conductor of electricity.	• Long antennas on vehicles may act as conductors. • Communicate power line locations for air resources. They may not be able to see them. • Avoid parking directly under power lines. • Power line right-of-ways should not be used as control lines if firefighters will be working directly under the power lines.

© Tom Oliveira/Shutterstock.com

Hazardous Materials (HazMat)

The likelihood of hazardous materials is increased with structures located in rural areas. People living in these areas may have more hazardous materials as part of their lives or livelihoods than people living in an urban area. All structures should be thoroughly evaluated for hazardous materials before protection measures are implemented. If a structure is on fire, then firefighters should consider the situation to include hazardous materials. Even if the structure does not have hazardous materials, it is better to consider that it does.

Hazardous materials should be handled by professionals; those who are trained, certified, and equipped to deal with the situation. If you locate any potential hazardous materials, flag the area and avoid exposure to firefighters. Contact your supervisor. Once you are in a safe location, deny entry and wait for hazmat professionals.

Unique hazardous materials often found in a wildland-urban interface fire include:

TABLE 16-2 *UNIQUE HAZARDOUS MATERIALS*

Propane Tanks	**Petroleum Tanks**
 © Tom Oliveira/Shutterstock.com	 © Steve Collender/ Shutterstock.com
Potential to BLEVE.	Liquefied Petroleum Gas (Propane)

(continued)

Flammable Liquids

© Kathy Clark/Shutterstock.com

Check for gas cans, fuel containers, etc.

Paint Cans

© Juhku/Shutterstock.com

Pesticides

© Jiggo_thekop/Shutterstock.com

Herbicides/Cleaning Materials

© Jeffrey B. Banke/Shutterstock.com

Fuel Tanks

© rthoma/Shutterstock.com

Drums

© isaravut/Shutterstock.com

Ammunition

© Kevin Brine/Shutterstock.com

Explosives

© Fer Gregory/Shutterstock.com

(continued)

Vehicles

Courtesy of Brian Henington

Plastics/Synthetic Material

© Gorich/Shutterstock.com

Illicit Labs

Illicit labs have the potential to kill firefighters. Illicit labs may include drugs, moonshine facilities, chemical agents, explosives, and/or biological labs. The most common lab that firefighters will encounter will be drug labs, with methamphetamine (meth) labs being the most common. The hazardous materials associated with these labs may cause great harm to firefighters. If the drug lab catches on fire, the risk to firefighters is greatly increased. "In illicit labs, both the final product and the production materials are harmful. These will vary depending on the type of lab, but one can expect a variety of flammable, toxic, and biological hazards."[5]

You may have been tasked with triaging and establishing structure protection measures on an unknown drug lab. You should pay attention to any suspicious activities or materials. Some indications of illicit labs are identified below. If you encounter a combination of these visual indicators, you should consider that the structure might be an illicit lab. Remove yourself and contact your supervisor.

Visual Indicators of Illicit Labs

- Large amounts of trash outside of the structure.
- Windows covered with plastic or tinfoil.
- Excessive propane tanks.
- Structures or pavements appear discolored.
- Attack dog breeds.
- Activities that make the structure hard to see.
- Strong odors of solvents, ammonia, starter fluid, or ether.

Producers of illegal drugs or other activities will protect their facilities from law enforcement and/or competitors. Firefighters should be aware that booby traps might be present at any illicit lab. Again, if you feel the structure is suspicious or the occupants are acting suspiciously, remove your team from the area and contact your supervisor who will advise the fire managers. Some examples of booby traps include:

[5] International Fire Service Training Association. *Hazardous Materials for First Responders: 4th Edition* (Stillwater, Oklahoma: Oklahoma State University, Board of Regents, 2010), 627.

- Fishing line with fishing hooks designed to hook and rip eyes
- Explosives such as dynamite or grenades
- Bottles filled with chemicals that will produce extreme toxins when broken
- Spikes or boards with nails sticking up
- Animal traps

Potential Hazards

Other hazards located in the interface may include the following: (These items may not be hazardous but do have the capability to become dangerous to firefighters.)

- Homeowners who refuse to evacuate (not properly trained and may act irresponsible, emotional, or threatening to firefighters)
- Homeless or abandoned people
- Pets left behind
- Livestock
- Media
- Traffic

TACTICAL PLANS

Fire managers utilize three primary tactical plans to combat wildland fires burning in the interface. The three plans include: 1) Defensive Mode, 2) Offensive Mode, and 3) Combined Mode. Large fires may utilize all three plans at any given time. The determining factor is based on the structures and expected fire behavior. The tactical mode should be based on an adequate examination of the situation. LCES should be in place and safety zones should be identified and adequate. If they are not, then no tactical mode may be acceptable. In addition, lookouts should also be posted to monitor fire behavior as it approaches your location.

The **defensive mode** can be described as protecting and prepping structures without any action to suppress the wildland fire. The defensive mode is often referred to as structure protection efforts. During this stage, firefighters may cut or remove flammable material next to the structure or remove flammable yard furniture. Defensive mode can also include establishing sprinkler systems to wet structures or wrapping structures with protective wrap. If a homeowner has developed and maintained defensible space, then the home may best suited for protection measures.

Certain circumstances may require fast or rapid protection measures. The decision to use rapid protection measures will be based on established and functioning LCES including adequate safety zones. Rapid protection measures may include closing windows or doors (leave unlocked), removing flammable material (including wood piles) away from the structure, and charging the home's garden hoses. The application of foam or fire gels can also be used during rapid structure protection measures.

The **offensive mode** involves the containment and control of a fire before it reaches structures. This tactical plan can involve fireline construction, burning out, or backfires. More than likely, firefighters will use a combination of defensive and offensive tactical plans. This is referred to as **combined mode**. The combined mode can involve structure protection measures and include fireline construction and burning out fuels around structures.

The tactical plans above must be constantly evaluated for firefighter safety. If the conditions worsen, the best option may include leaving the structure and taking refuge in a safety zone. "Regardless of the values at risk, safety for life is most important and is adhered to be the use of the risk management process, Standard Firefighting Orders, Watch Out Situations, and applying LCES."[6] The decision to withdraw from a fire should be based on the following:

1. Safety of your crew is in jeopardy.
2. Fire activity is intensifying and burning at extreme rates.
3. Numerous spot fires are occurring.
4. No radio or poor radio communication with supporting resources.
5. Emergency vehicles cannot point to their escape routes.
6. Inadequate water supply.
7. The roof of the structure is more than ¼ involved in fire.
8. The interior of structure is burning.
9. High, windy conditions exist or are anticipated.
10. Numerous structures are involved in fire at the same time.

Potential Water Sources

Firefighters may have to be creative when locating potential water sources in the interface. If water is readily available, then water delivery systems can be created that provide consistent water to structure protection efforts. If water is not available, firefighters should look for unique sources from which to draft or pull water. Some examples of effective water sources used on interface fires include:

- Swimming pools
- Hot tubs
- Storage tanks
- Fish ponds
- Irrigation ditches
- Dry hydrants
- Hydrants (Wildland engines should release pressure before connecting to hydrants. Some hydrants have high flow rate and gallon per minute outputs.)
- Standpipes

STRUCTURAL TRIAGE

Firefighters incorporate a process to evaluate structures and determine what protection tactical plan is warranted. This system is called structural triage. Triage allows firefighters to establish priorities based on which structures can be protected from advancing wildfires. The triage process allows firefighters to evaluate structure construction materials, adjacent fuels, ingress/egress, available resources, and fire behavior. Triage also utilizes the wildland-urban interface watch out situations and should be based on

[6] National Wildfire Coordinating Group. *Firefighter Training*, 14.3

LCES and the Rules of Engagement in the decision making process. Above all, triage considers firefighter exposure and safety as the ultimate determining factors.

Prior to any tactical decision, firefighters must ensure the following safety considerations are in place. They include:

1. Is a safety zone present and is it acceptable?
2. Is your vehicle pointed toward your escape route at all times?
3. Are you mobile and flexible? Can you exit the area very quickly if you have to?

These considerations should be in place every time you triage a structure. Each structure will require a new triage evaluation. Do not assume your evaluation of one structure will be acceptable for an additional structure. You need to be able to quickly retreat if the conditions become worse. Remember the following statement each time you begin the triage process: "Firefighters must know when it is time to pull back to a safety zone."[7]

The structure triage categories are identified below. These categories are also included in the *Incident Response Pocket Guide* and the *Wildland Fire Incident Field Management Guide.*

DEFENSIBLE—STAND ALONE

 # 1 Consideration: Are safety zones adequate and identified?

LCES—Ensure LCES is in place and functional!

Structure Characteristics:

- Structure has the ability to withstand fire activity with little or no assistance.
- Building construction: Stucco, bricks, metal roofs, Spanish tiles, etc.
- Surrounding vegetation—Is it flammable?
- Adequate defensible space.
- Good ingress/egress and turnaround spots.
- Is the structure part of a Firewise Community?

Courtesy of Brian Henington

The structure above needed no protection. The siding was stucco with a Spanish tile roof. Surface fuels were green mowed grass.

Tactics:

- Has few tactical challenges.
- Remove any flammable material (if any).
- Place lawn furniture in structure or far away from the structure.
- May have adequate water sources.

[7] Ibid., 14.11.

Patrol Efforts:

- Patrol after fire front passes.
- Look for spot fires and surface fires. Suppress immediately.
- If structure is on fire, stay away and allow structural firefighters to suppress the structure fire.

DEFENSIBLE—PREP AND HOLD

 # 1 Consideration: Are safety zones adequate and identified?

LCES—Ensure LCES is in place and functional!

Structure Characteristics:

- Structure needs preparation activities to help protect it.
- Building construction: Determine siding/roof. Metal roofs will improve structure's ability to withstand a fire.
- Surrounding vegetation—Is it flammable?
- How is the defensible space?
- May have minimal access issues.

Courtesy of Brian Henington

Firefighters covering entry routes for fire brands with structural wrap.

Tactics:

- May have some tactical challenges.
- Remove flammable material (if any).
- Cut and remove flammable material (with homeowner's permission)
- Place lawn furniture in structure or far away from the structure.
- Structural wrap may be used.
- May or may not have adequate water sources.
- Firefighters typically have to be on site to implement tactics as the fire front hits.

Courtesy of Brian Henington

Structural wrap on a propane tank.

Patrol Efforts:

- Patrol after fire front passes.
- Look for spot fires and surface fires. Suppress immediately.
- If structure is on fire, stay away and allow structural firefighters to suppress the structure fire.

NON-DEFENSIBLE—PREP AND LEAVE

 # 1 Consideration: SAFETY ZONES ARE NOT ADEQUATE OR AVAILABLE!

LCES may not be able to be implemented.

Structure Characteristics:

Courtesy of Brian Henington

- Structure needs preparation activities to help protect it, but such activities may have to be performed quickly.
- Building construction may be conducive to combustion.
- Surrounding vegetation may be flammable.
- Limited defensible space.
- May have access issues and inadequate turnaround spots.

Tactics:

- Has tactical challenges.
- Crews cannot commit to be on site when fire front hits the structure.
- Rapid mitigation measures may be the best option.
- Remove any flammable material.
- Place lawn furniture in structure or far away from the structure.
- May or may not have adequate water sources.
- Have the homeowners done their part to protect their home prior to a fire ever occurring?

Patrol Efforts:

- Patrol after fire front passes.
- Look for spot fires and surface fires. Suppress immediately.
- If structure is on fire, stay away and allow structural firefighters to suppress the structure fire.

NON-DEFENSIBLE—RESCUE DRIVE-BY

 # 1 Consideration: SAFETY ZONES ARE NOT PRESENT!

LCES is not in place or adequate.

Structure Characteristics:

© Steve Photography/ Shutterstock.com

- Structure is expected to burn.
- Building construction is conducive to combustion.
- Surrounding vegetation is flammable.
- Zero defensible space.
- May have access issues and inadequate turnaround spots.

Tactics:

- Significant Tactical Challenges!
- Crews cannot commit to be on site when fire front hits the structure.
- If there is time, check to ensure no civilians are in the structure.
- Trigger points should be established. Leave the area when the fire hits a certain geographical point.

Patrol Efforts:

- Patrol after fire front passes.
- Look for spot fires, surface fires. Suppress immediately.
- If structure is on fire, stay away and allow structural firefighters to suppress the structure fire.

WILDLAND-URBAN INTERFACE WATCH OUT SITUATIONS

In addition to our Rules of Engagement, fire managers have created specific guidelines for the wildland-urban interface. These guidelines are called: The Wildland-Urban Interface Watch Out Situations. These guidelines have been in place for several years; however, the National Wildfire Coordinating Group has published a new set of guidelines. These guidelines, as of 2014, are located in the *Wildland Fire Incident Management Field Guide—2014 Edition*.

1. Poor access and narrow, one-way roads

   Concerns:

   » Narrow roads may not be safe for vehicles.
   » One-way traffic (ingress/egress may be compromised).
   » Inadequate road construction.
   » Dead ends or small cul-de-sacs.
   » Vehicles may not be able to turn around.
   » Not able to see approaching traffic.
   » Evacuating public may hinder emergency vehicle traffic.
   » Fuel arraignments along road may compromise firefighter safety.
   » Roads on slopes can be highly impacted by extreme fire behavior. Not a safe location for firefighters.
   » Type 1 or 2 engines may have difficulty accessing structures.
   » Access should be restricted. Use law enforcement support.

© wavebreakmedia/Shutterstock.com

Courtesy of Brian Henington

2. Bridge load limits

   Concerns:

   » Inadequate construction techniques and materials.

   » May not support Type 1 or Type 2 fire engines because of weight limits.

   » Heavy equipment may not be able to use bridges.

   » Always inspect a bridge prior to traveling across it.

© Sandra Kemppainen/Shutterstock.com

3. Wooden construction and wood shake roofs

   Concerns:

   » Dry wood siding and roof shakes are highly susceptible to ignition.

   » Radiant heat from advancing fire or adjacent involved structures may preheat the structure and allow ignition of the structure.

   » Pine needles, leaves, or other flammable debris on the roof or next to the structure will increase the chances of the home burning.

   » Hard to defend if defensible space is not in place.

© Shaiith/Shutterstock.com

4. Power lines, propane tanks, and Hazmat threats

   Concerns:

   » Increased exposure to firefighters.

   » Firefighters should expect any or all three of these factors to be in play.

   » Downed or damaged power lines may be energized.

   » Hazmat situations may be numerous, and very dangerous to firefighters.

   » Propane tanks may BLEVE.

   » If any of these hazards are in play, the best mitigation measure may be complete avoidance.

Courtesy of Brian Henington

5. Inadequate water supply

   Concerns:

   » Limited water supplies.

   » Multiple fire resources using the same water supply.

   » May not have fire hydrants.

   » Type 1 or Type 2 engines may not be effective without adequate water.

   » Preserve your water. Always keep a minimum of 100 gallons of reserve water in your engine for protection of your engine crew.

   » Remain as mobile as possible.

   » Avoid long hose lays that are connected to engines.

   » Avoid wetting roofs/structures or surface fuels if the fire is far away.

Courtesy of Brian Henington

6. Natural fuels 30 feet or closer to structures

   Concerns:

   » Preheating of structure by adjacent burning fuels.

   » Defensible space. Has the homeowner created defensible space? If not, do you have enough time to do so?

   » Firewood stacked next to home or under wood decks.

   » Homes located on steep slopes must have a greater defensible space on the side facing the bottom of the slope.

   » Prepping the home may require removing firewood, cutting flammable material, and removing surface fuels.

   » Homeowners may not allow firefighters to cut trees or debris. Always document your activities and ask permission from the homeowner or fire manager.

Courtesy of Brian Henington

Courtesy of Brian Henington

7. Structures in chimneys, box canyons, narrow canyons, or on steep slopes (grade 30% or more)

   Concerns:

   Homes in these conditions have the lowest survivability and present the highest safety risk to firefighting personnel.[8]

   » Fire may release all of its energy at the structure.

   » Wind is channeled and highly influenced by these terrain features.

   » Extreme burning indices and heat release rates.

   » Extreme rates of spread.

   » Firefighter safety will be compromised.

   » LCES may be compromised.

© Tom Reichner/Shutterstock.com

8. Extreme fire behavior

   Concerns:

   "This behavior has been responsible for major losses of life and property within the wildland-urban interface"[9]

   » Monitor fuel and weather conditions that may elevate fire behavior and create extreme conditions.

   » Understand what the atmospheric conditions are.

   » Consider the burn period and the potential for elevated fire behavior.

   » If topography (terrain) is aligned with wind and fuels, expect extreme fire behavior.

---

[8] Ibid., 14.13

[9] Ibid., 14.9

» Extreme fire behavior may limit the effectiveness of fire suppression activities.

» Pay attention to any of the following. They are an indication of extreme fire behavior.

- Spot fires
- Torching
- Crown fires
- Fire whirls
- High rates of spread

9. Strong winds

Concerns:

» Highly influence fire behavior.

» Monitor wind conditions at all times.

» Surface winds over 10 mph will have a major impact on fire behavior.

» A major factor in crown fire development.

» Firebrand transportation.

» Increased oxygen available for the fire.

» Monitor weather events such as cold fronts, thunderstorms, foehn winds.

© Kaleidoscopio/Shutterstock.com

© Jenn Huls/Shutterstock.com

© Arthur Eugene/Shutterstock.com

10. Evacuation of public (panic)

Concerns:

» Fire managers or law enforcement may order evacuations to help protect the public from the approaching fire threat.

» Evacuations may also assist firefighter activities by mitigating the concern of the public interfering with their operations.

» Whenever the public is involved, the complexity, confusion, and stress levels of firefighters will greatly increase—which may result in rude or offensive behavior by the public. Do not engage an aggressive individual; request law enforcement to deal with the issue.

» Law enforcement should be responsible for evacuation efforts.

» Fast moving fires may impact an effective evacuation.

» Law enforcement may not be trained, equipped, or prepared for the fire activity. If you feel their safety is in jeopardy, politely ask them to leave for their own safety.

11. Underground utilities threat

Concerns:

- May include electricity, natural gas, propane, and/or other utilities that may impact firefighter safety.
- Heavy equipment rupturing underground utilities.
- Release of natural gas or propane.
- Fire activities may cause damage to underground utilities that may cause additional hazards to firefighters.
- Fire managers should request assistance from utility companies to identify issues.
- Utilities may be shut off/closed to reduce additional threats to firefighters.

12. Structural collapse zone when structures are exposed to fire

Concerns:

- Highly dangerous to firefighters and the public.
- Building collapses have been associated with firefighter injuries and fatalities.
- If a roof is ¼ involved with fire, remove your team from the area. "The situation is usually hopeless."[10]
- Windows are black, sweating, or bulging; back-draft conditions may be present. Remove your team from the collapse zone.

© Knumina Studios/Shutterstock.com

13. Smoke byproducts often laced with chemical compounds not found in pure wildland fires

Concerns:

- In general, wildland firefighters do not have devices to protect their airways.
- Stay upwind of a burning structure.
- Treat as a hazardous materials situation.
- Remove your team far enough away from the burning structure to avoid inhaling these extremely dangerous byproducts.

[10] Ibid., 14.5.

FIREFIGHTER COMPASSION

Our main responsibility in interface fires is the safety of firefighters. The second priority is the protection of the public. The best protection method to ensure the public safety is evacuations. Evacuations are intended to protect the public from the dangerous conditions present with an interface fire. Evacuations also are intended to reduce the potential for emotional or hostile encounters with emergency responders. If an evacuation is ordered, firefighters should be aware of the stress the homeowners are experiencing. We should be compassionate to their situation.

This author has been involved in numerous evacuations. Some have been small; some have included entire cities. I remember one evacuation that involved the entire community of Los Alamos, New Mexico. The evacuation was ordered for 10,000+/- people. Although evacuation centers were established, there were not enough. Many people had to stay at motels. The motels gouged their nightly prices, which accelerated stress and anger levels among the evacuees. Homeowners would spend the majority of their time watching the news and begging for any available information regarding their homes.

I remember another time when a rancher was forced to evacuate his ranch. He had 350 yearling cattle in a corral system. He was not allowed to return to the ranch to water or feed the cattle after he was evacuated. His livelihood was at risk. This, too, caused considerable stress to the rancher, which triggered an angry and emotional outburst.

Regardless where the evacuation occurs or how many people are involved, we must understand that people have left behind something that means a lot to them. They are under a great deal of stress. We should take every measure to ensure our firefighting activities do not damage, deface, and/or destroy their personal property.

Matthew Pena (Fire Science Instructor) identifies the necessity of remaining considerate for homeowner's property and the issues we have in this very stressful situation:

> *It is important for us as firefighters to remain sensitive and open-minded while working in the wildland-urban interface. We are dealing with people's personal property and homes, emotions are running very high, and there is a high expectation of us as firefighters.*

SUMMARY

The wildland-urban interface presents unique and unfamiliar challenges for wildland firefighters. Become familiar with the concepts discussed in this chapter to help build your situational awareness of these hazards and the appropriate mitigation measures. LCES is critical in the interface and should always be functioning and constantly reevaluated throughout the operational period. Finally, no structure is more important than your life or the lives of your fellow firefighters. Only take action when it has been determined that you can do so safely.

KNOWLEDGE ASSESSMENT

1. Define the wildland-urban interface.
2. True/False: A firefighter should do everything possible to protect structures at all costs.
3. Identify five unique hazardous materials often found in the interface.
4. What labs are considered illicit?
5. Identify the three tactical plans used in interface fires.
6. What are four potential water sources used in interface fires?
7. Define "structure triage."
8. What is the ultimate determining factor for triage?
9. Identify the four categories of structural triage.
10. True/False: Firefighters should not be considerate to home owners property and belongings?

EXERCISES

1. Identify and explain the safety concerns with downed power lines.
2. Research the term "defensible space." What does the term mean? How does it support triage decisions. List your reference(s).
3. Research Firewise Communities: What are they? How do Firewise Communities help firefighters protect structures? List your reference(s).

BIBLIOGRAPHY

International Fire Service Training Association. *Hazardous Materials for First Responders: 4th Edition*. Stillwater, Oklahoma: Oklahoma State University, Board of Regents, 2010.

National Wildfire Coordinating Group. *Firefighter Training, S-130*. Boise: National Wildfire Coordinating Group, 2003.

National Wildfire Coordinating Group. *Fireline Handbook*. Boise: National Wildfire Coordinating Group, 2004.

National Wildfire Coordinating Group. *Glossary of Wildland Fire Terminology*. Boise: National Wildfire Coordinating Group, 2014.

National Wildfire Coordinating Group. *Incident Response Pocket Guide—2104* Version. Boise: National Wildfire Coordinating Group, 2014.

National Wildfire Coordinating Group. *Wildland Fire Incident Field Management Guide: 2014*. Boise: National Wildfire Coordinating Group, 2014.

Pena, Matthew (Fire Science Instructor, Central New Mexico Community College, Albuquerque, New Mexico). Interview with author, March 20, 2015.

CHAPTER 17

BASIC LAND NAVIGATION

Learning Outcomes

- Explain the importance of maps as they relate to wildland fire suppression.
- List the key features of maps.
- Discuss the Public Land Survey System and how it is used in fire management activities.
- Apply the basic skills necessary to effectively use a compass on a wildland fire.
- Explain the latitude/longitude mapping system and how it relates to wildland fire activities.
- Identify the key ICS mapping symbols used on wildland fires

Courtesy of Brian Henington

Key Terms

Azimuth
Compass
Contour Intervals
Contour Lines
Declination
Global Positioning System (GPS)
Index Contour Lines
Latitude and Longitude
Magnetic North
Maps
Public Land Survey System
True North

OVERVIEW

The ability to effectively use a map is an essential skill for wildland firefighters. This is often a skill that is not emphasized or prioritized by some instructors. The misconception that instructors have is that map reading should be the responsibility of the student. This is the wrong approach. There are too many firefighters of all experience levels who cannot effectively read a map or who count on somebody else to do it for them.

The purpose of this chapter is to introduce you to the basic navigation skills needed for a wildland firefighter. You will be introduced to fire maps, map features, the Public Land Survey System, compass use, latitude/longitude, and ICS mapping symbols.

FIRE MAPS

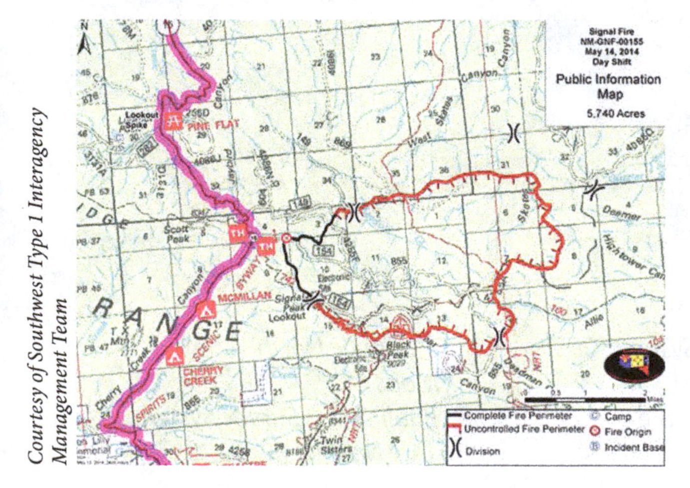

Courtesy of Southwest Type 1 Interagency Management Team

Taken from Southwest Area Type 1 Interagency Management Team.

A map is a reflection of the surface of the earth. According to the National Wildfire Coordinating Group, "A map is a graphic representation of the earth's surface and is similar to an aerial view of a portion of the earth."[1] As with all fire suppression concepts, all actions are based on the safety of firefighters. Not being able to use a map can, and has, jeopardized the safety of firefighters.

Maps are used on wildland fires for all aspects of fire management. We use them for operational planning, operational engagement, transportation routes, navigation, incident status, and public information. Maps also provide a function that facilitates decision making and hazard avoidance. Maps can specifically support the following activities:

- Assist with navigation efforts
- Calculate distance
- Identify vegetation cover
- Show routes of travel or access points
- Calculate the size of a fire or specific area
- Identify the location of a specific point of interest or concern
- Provide a visual overview of a specific area.
- Identify terrain features that influence fire behavior
- Identify names of rivers, creeks, mountains, peaks, roads, etc.

[1] National Wildfire Coordinating Group, *Firefighter Training*, S-130 (Boise: National Wildfire Coordinating Group, 2003), 16.2

Fire managers will utilize specific maps to assist with fire management activities. Some examples are included below:

- Topographic
- Subdivision
- Road (Highway)
- Incident Action Plan
- Facilities
- Infrared
- Land Ownership
- Operational Briefing
- Public Information
- Situation
- Fire Progression
- Orthoimagery

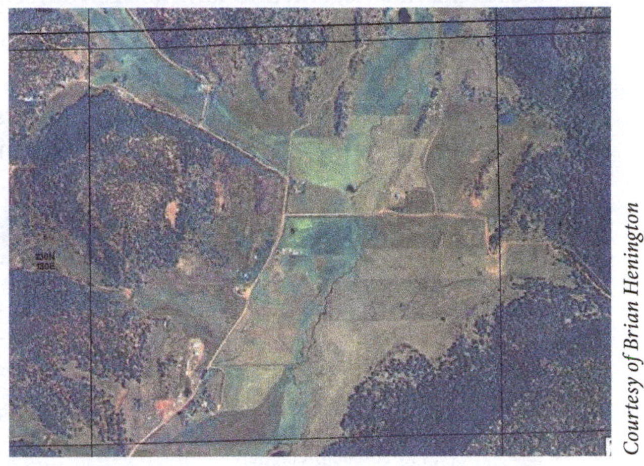

Courtesy of Brian Henington

Example of an Orthoimagery Map.

MAP FEATURES

The effective use of maps requires comprehension of specific map features. The key features of common fire maps include the following:

- North arrow: Points toward the top of the map (typically). Always check for a north arrow on a map before use. Some maps are photocopied. They may not have a north arrow identified on them.
- Map scale: A representation of a measurement on the map as it relates to the measurement on the ground or surface. A map that has a 1:24,000 scale indicates that one inch on the map would reflect 24,000 inches on the ground (about 2,000 feet). Scales are usually located at the bottom of the map.
- Map name: Maps will typically be assigned a name. Located in the top right corner or bottom right corner of the map.
- Counter interval: Identifies the vertical difference between each contour line. The interval of the individual map should be located on the bottom of the map. For example, 40 feet, 50 feet, etc.
- Quad size: Standard location is the upper right corner of the map. Provides the minutes of latitude contained in the map. Examples include: 7.5 and 15 minute quads. 7.5 quads = a map scale of 1:24,000.
- Declination: Standard location is at bottom of the map. Check the map's declination and apply the appropriate declination setting to your compass.
- Other scales: May include latitude/longitude scales, map datum, UTM, or others.
- Road Classification: Identifies the type of roads located on the map. Standard location is the bottom right corner of the map.

- Map legend: Can be located on the bottom of the map or on the side of the map. Provides a complete description of symbols, color indicators, land ownership, and/or may include roads. Always locate the map legend and familiarize yourself with the features before studying the map.
- Map symbols: Identifies features located on the map. Examples include water features, vegetation, elevation, human-made, etc.
- Map colors: Check the legend, some producers of maps may use different color schemes to identify certain features. Standard colors on fire maps are identified below.

Blue	Facilities, water	**Green**	Vegetation
Red	Fire features, roads, origin	**Brown**	Contours, cuts/fills, other relief features
Black	Roads, control lines, drop points	**Purple**	Revised information
Orange	Fire spread prediction	**Grey**	Developed areas

Contour Lines and Intervals

Contour lines are components featured on topographic maps. They can also be located on other types of maps, but the main use is to interpret elevation changes on topographic maps. Contour lines are identified as, "actual lines on a map along which every point is at the same height above sea level."[2] Contour lines will start at one point on a map and close at the same point.

Contour lines will help firefighters locate terrain influences that may influence fire behavior and jeopardize firefighter safety. Some common terrain features include:

Courtesy of Brian Henington

[2] Ibid., 16.6

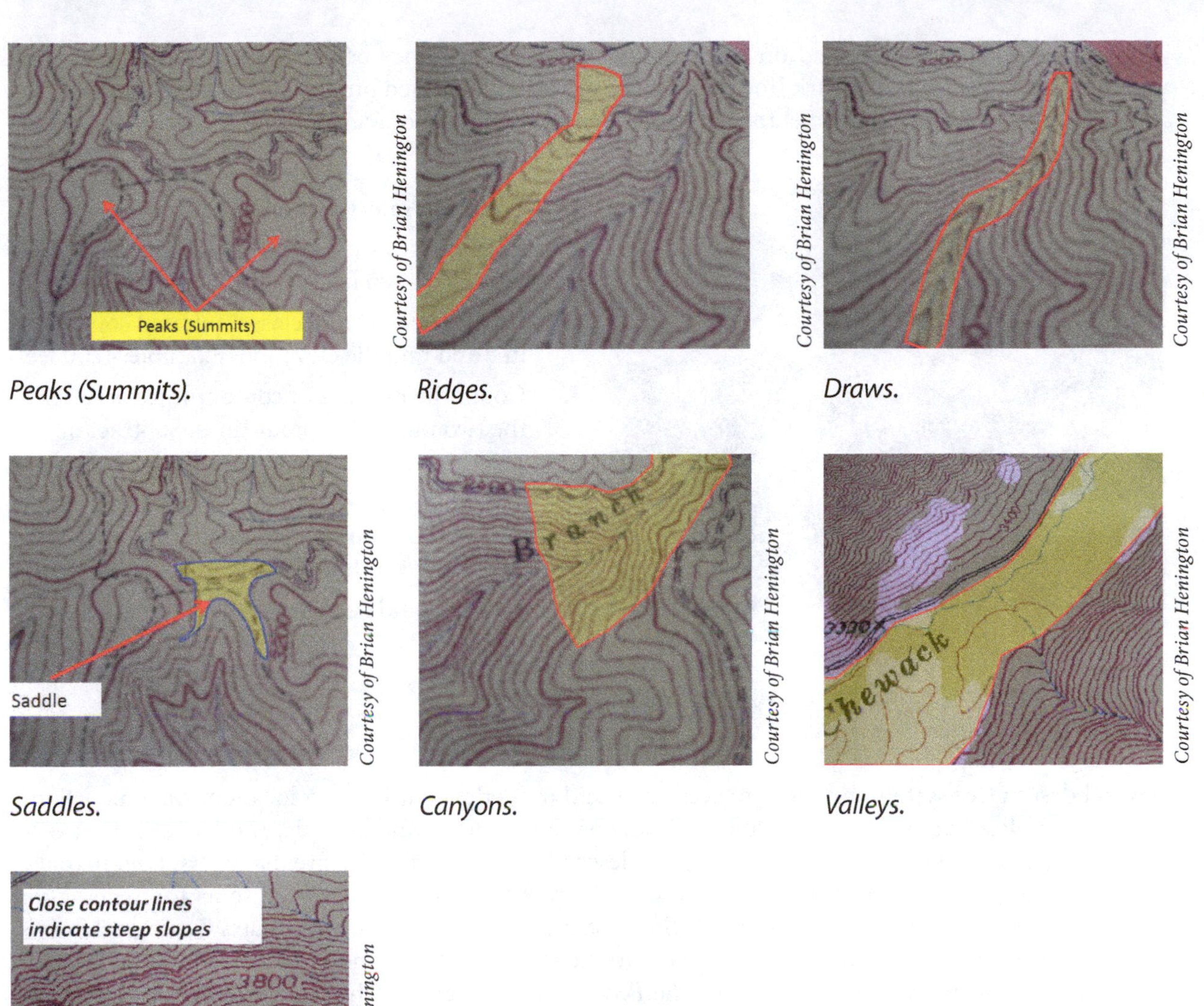

Courtesy of Brian Henington

Peaks (Summits).

Ridges.

Draws.

Saddles.

Canyons.

Valleys.

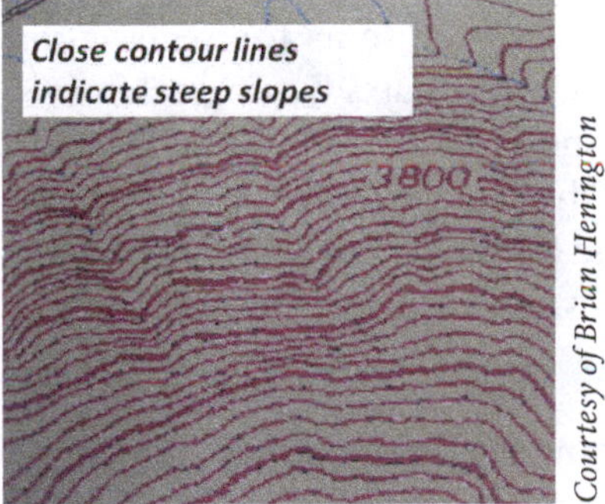

Courtesy of Brian Henington

Steep Slopes.

In addition, contour lines will also indicate terrain features that should be also considered by firefighters.

- V: contour lines creating a V indicate a water feature. The V will point upstream.
- U: identify a ridge. The U will point down the ridge.
- Closed contour lines: indicate a summit (peak).
- Closed contour lines with tick marks: indicates a depression.
- Waterfalls, overhanging cliffs: contour lines will cross or meet.
- Close or tight contour lines: steep slopes.
- Wide or loose contour lines: flat terrain or gradual slope.
- Irregular contour lines: Indicate rugged and rough terrain. Firefighters should avoid if possible.

Contour intervals are identified on the map and depict the distance between each contour line. The index contour line will be a dark line that has the elevation identified on it. You can also use the index contour line to determine the distance of the contour lines if it is not identified on the map. To do this, follow the steps below:

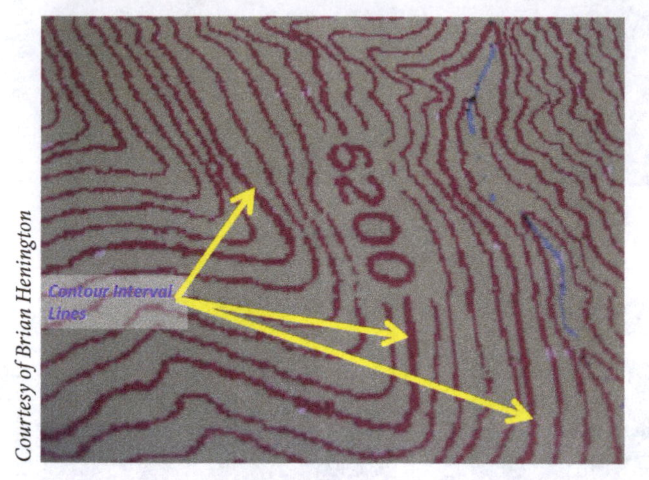

Courtesy of Brian Henington

1. Locate an interval contour line.
2. Find the elevation and write it down. For example, 4400 feet.
3. Locate the next and closest interval contour line and find the elevation. Example: 4200 feet.
4. Count the number of contour lines between the two interval contour lines. Subtract or add the first interval from or to the second. 4400 — 4200 = 200
5. Divide the result by the number of contour lines between the intervals. 200/5 = 40.
6. 40 feet would be the contour interval.

PUBLIC LAND SURVEY SYSTEM

Several descriptive systems are used to accurately and properly identify exact locations on a map. Some examples of descriptive systems include: latitude/longitude, metes and bounds, and Universal Traverse Mercator (UTM). The most common mapping descriptive system used by fire managers, land management agencies, and surveyors is the Public Land Survey System (often referred to as the rectangular survey system or legal land survey system). The PLSS is explained by the United States Geological Survey as "...a way of subdividing and describing land in the United States... The PLSS is used to divide public domain lands, which are lands owned by the Federal government for the benefit of the citizens of the United States."[3] The system was not enacted until 1796 by the Public Land Survey law; therefore not every state is surveyed under this system. Today, there are 30 southern and western states that use the PLSS as the primary survey system to identify locations on a map.

The PLSS produces a legal land description. This information is important to pinpoint an exact location on a map or an area. The PLSS utilizes the concepts of Township, Range, and Section(s) to identify locations on the ground. The key terms and concepts of the PLSS are identified below.

[3] United States Geological Survey, *The Public Land Survey System (PLSS)*, accessed April 1, 2015, http://nationalmap.gov/small_scale/a_plss.html#one.

Term	Definition	Visual Reference
Initial Point	Each state has its own initial point and many large states may have more than one initial point. The initial point establishes where the base line and principal meridian initiate.	Courtesy of Brian Henington
Base Line	Base line runs east/west and is intended to divide townships into north or south.	Courtesy of Brian Henington
Principal Meridian	Principal meridian runs north/south and is intended to divide townships into ranges that run east or west.	Courtesy of Brian Henington
Township	Townships are an essential component of the PLSS. A township consists of 36 sections, or is 6 miles x 6 miles (36 square miles). Townships are established from the base line. Any township north of the base line is considered a north township. Any township south of the base line is a south township. Township lines run east/west. Townships are identified as follows: Township 1 North or T1N. You can locate the township on the left or right side of the map.	Courtesy of Brian Henington When identifying a location on a map, the township is always listed before the range. Township 24 North, Range 16 East.

(continued)

Technique	Process	Visual Reference
Range	A range is as important as a township. A range runs north/south and divides townships into east or west designations. The range will begin at the principal meridian and will be identified as either west or east. Ranges are identified as follows: Range 6 East or R6E. Range numbers are located on the top and bottom of the map. When identifying a location on a map, the range is always listed after the township. Township 24 North, Range 16 East.	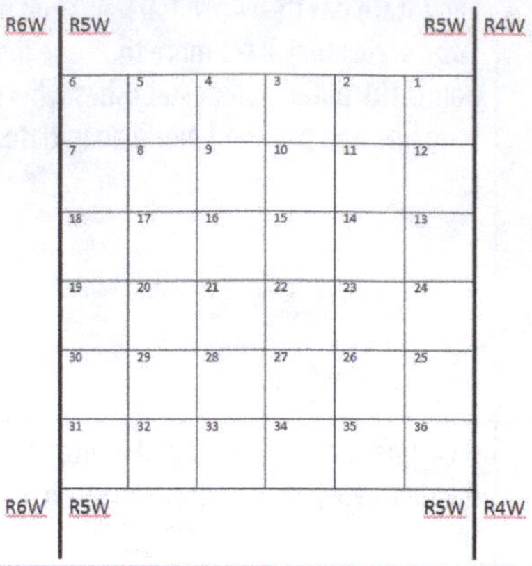 Courtesy of Brian Henington
Section	Townships are divided into 36 sections. A section was intended to be one square mile (80 chains x 80 chains). Furthermore, a section consists of 640 acres. Sections are identified by the number. Example here is section 16. Note: Some townships have been corrected due to more accurate technology. You may encounter a section that is not 640 acres; it could be more or less.	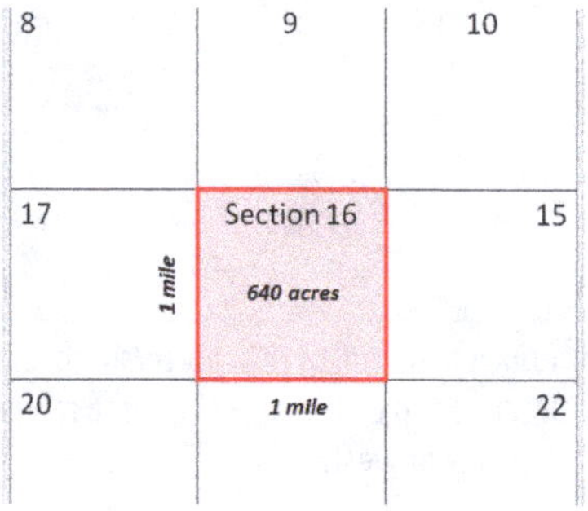 Courtesy of Brian Henington
Half (1/2) Section	A half section is 320 acres or ½ mile x 1 mile. Half sections are determined by dividing the section into north and south, across the middle of the section. Maps will typically not have a half section identified, but firefighters must know how to divide the section into half sections. Half sections are identified as: north half of section 16 or south half of section 16.	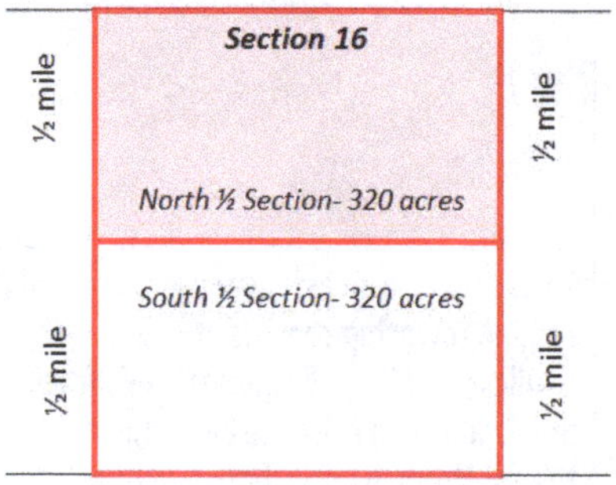 Courtesy of Brian Henington
Quarter (¼) Section	A quarter section is 160 acres or ¼ mile x ¼ mile. Quarter sections are determined by dividing the section into four quarters. Divide the section into north/south and east/west. Just as the case of a half section, firefighters must know how to divide the section into quarters. Quarter sections are identified as: northwest, northeast, southeast, or southwest quarter. If you are using quarters to identify a location of a map, you do not have to use the half section identifier.	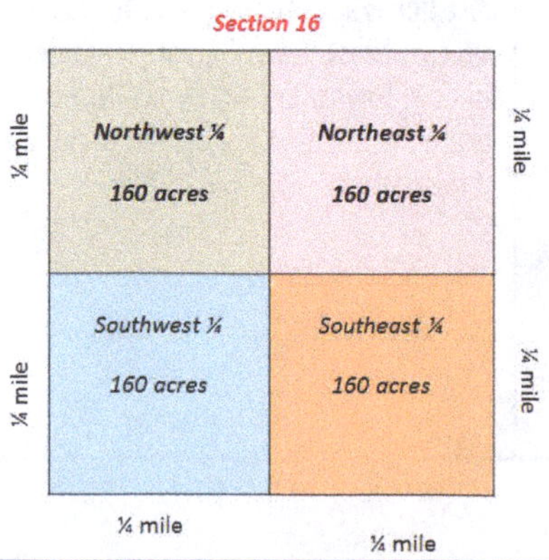 Courtesy of Brian Henington

(continued)

Technique	Process	Visual Reference
Quarter/ Quarter Section	Quarter/Quarter sections are used to identify a location into 40 acres. Quarter/quarter sections are designed to further pinpoint an exact location on a map. A section is identified into 16 quarter/quarter sections. The proper way to identify quarter/quarter sections are: The northwest quarter of the northwest quarter or NW1/4 NW 1/4. The northwest quarter would identify that this quarter belongs to the northwest quarter. Quarter/quarters are not normally identified on maps. Firefighters should still be able to locate an exact location by using the quarter/quarter identification system.	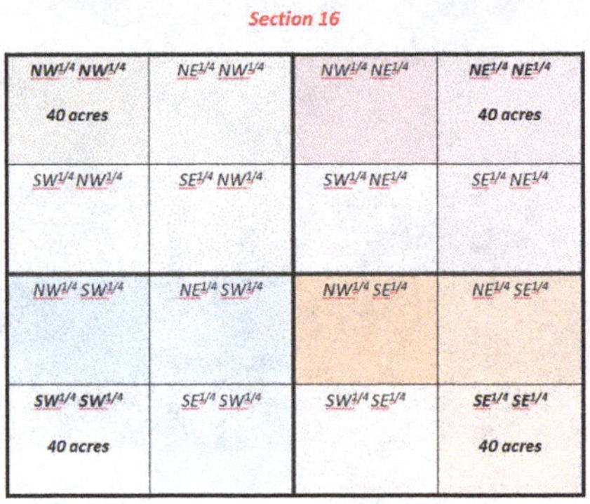 Section 16 NW1/4 NW1/4 40 acres · NE1/4 NW1/4 · NW1/4 NE1/4 · NE1/4 NE1/4 40 acres SW1/4 NW1/4 · SE1/4 NW1/4 · SW1/4 NE1/4 · SE1/4 NE1/4 NW1/4 SW1/4 · NE1/4 SW1/4 · NW1/4 SE1/4 · NE1/4 SE1/4 SW1/4 SW1/4 40 acres · SE1/4 SW1/4 · SW1/4 SE1/4 · SE1/4 SE1/4 40 acres *Courtesy of Brian Henington*
Acre	An acre consists of 43,560 square feet. An acre is about the size of a football field, if the football field were square. There are 640 acres in a section, 320 acres in a half section, 160 acres in a quarter section and 40 acres in a quarter/quarter section.	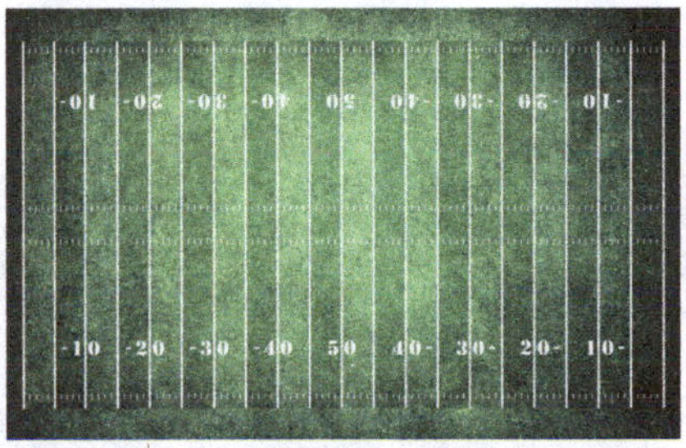 *© maodoltee/Shutterstock.com*

Numbering Sections

Section numbering begins in the top right corner (Section # 1) of a township and ends with section 36 in the bottom right corner. Many maps will only identify section 2, 16, 32, and 36. The illustration below is the section numbering system of a township.

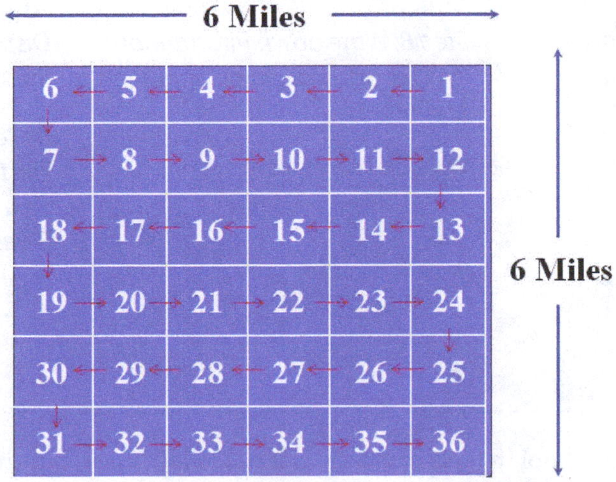

6 Miles (width) × 6 Miles (height)

6	5	4	3	2	1
7	8	9	10	11	12
18	17	16	15	14	13
19	20	21	22	23	24
30	29	28	27	26	25
31	32	33	34	35	36

Courtesy of Brian Henington

Identifying Legal Land Descriptions

Identifying the legal land description of a location is included in the steps below.

Courtesy of Brian Henington

Step 1: Locate the fire or destination on the map.

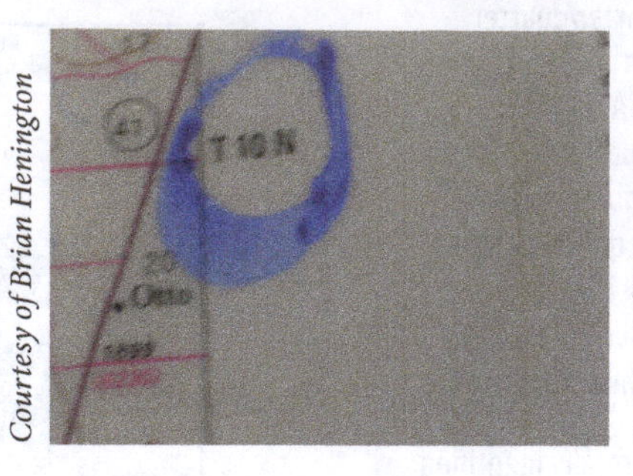

Courtesy of Brian Henington

Step 2: Locate the township on right/left side of map. Write down on a piece of paper.

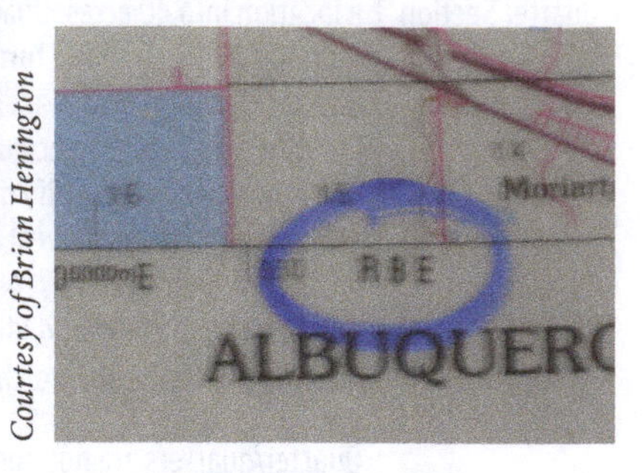

Courtesy of Brian Henington

Step 3: Locate range on the bottom/top of the map. Write down on a piece of paper.

Courtesy of Brian Henington

Step 4: Match township and range by intersecting their location.

Courtesy of Brian Henington

Step 5: Identify the section.

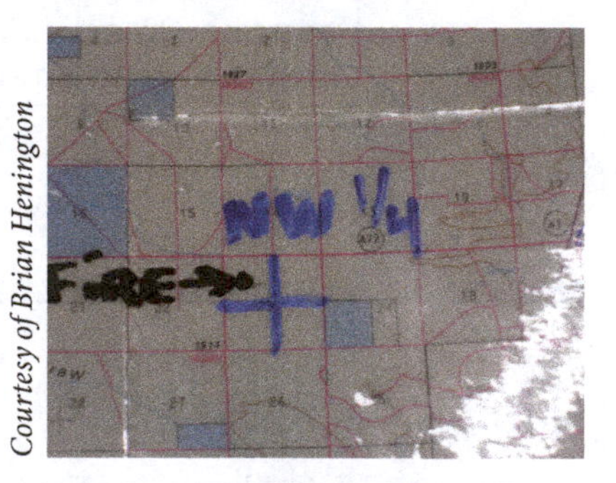

Courtesy of Brian Henington

Step 6: Bisect section into quarter sections to determine location of the target area.

Step 7: Further bisect quarter sections into quarter/quarter sections.

Step 8: Write down information and report to dispatch or other firefighters.

Dispatch stand by to copy location of the fire:

Bailee Fire is located in the Northwest ¼ of Section 23, Township 10 North/Range 8 East.

THE COMPASS

The compass is an essential tool in land navigation. It has several uses, but the most common is to determine direction. Firefighters use the compass to determine cardinal directions and wind direction as well as to determine routes of travel. This section will provide a brief overview of a compass and its functions. As in the case of most tools, practice with the compass will improve your ability to effectively use one.

Orienting Arrow: The orienting arrow allows the user to align the magnetic needle when taking an azimuth/bearing. Magnetic declination can also be set using the orienting arrow.

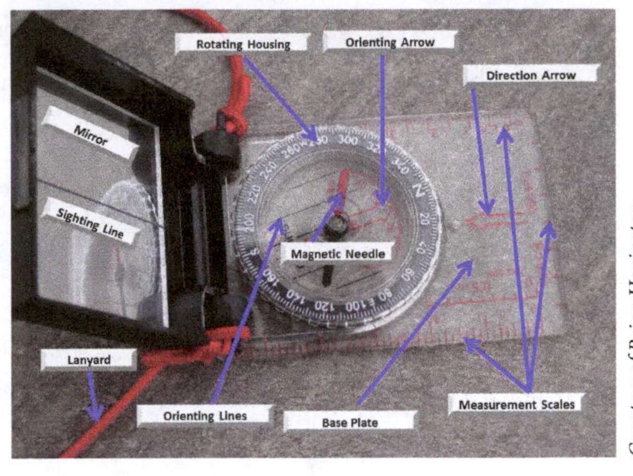

Courtesy of Brian Henington

Measurement scales: Provides measurements scales (varies per compass). Assists in determining distances on a map.

Base Plate: Used as a platform for the compass and has measurement scales to assist in navigation.

Direction Arrow: Identifies the direction of travel. When assisting navigation efforts, the arrow should be pointed toward the destination.

- Dial with Degrees (Rotating Housing): allows user to spin the housing unit to match the cardinal direction (north, east, south, or west). Degrees are identified 0-360°.
- Magnetic Needle with Red North: Located inside the rotating dial. Pivots and rotates to determine magnetic north.
- Orienting Lines: Parallel to the orienting arrow. Used with grid lines on a map. Located inside dial (rotating housing).
- Lanyard: Holds compass. Can be worn around the firefighter's neck or attached to other equipment.
- Declination Adjustment: Allows user to adjust compass to the declination of a specific area. Declinations can be located on most maps.
- Mirror: Used to improve azmiuth/bearning readings. Can also be used as communication tool to identify your location to other firefighters or aircraft.
- Cover: Protects the compass from damage.

Declination Setting

Firefighters must know the declination setting of the geographic location where you are fighting fire. To do this, you must first understand the difference between magnetic north and true north. True north can be explained as the direction in relation to the North Pole. Maps are created based on true north. Magnetic north deals specifically with the compass. Magnetic north is the direction the red arrow on a compass will determine as north. The difference between true north and magnetic north is called declination.

Declination settings are different based on your location in the United States. Declination settings are identified in degrees and in east or west of true north. Several online references can assist you with finding the declination of a specific area. Most maps produced by the federal government will also identify declination settings on the bottom of the map.

The type and make of a compass will determine how the declination is set. You should refer to the user guide for your specific compass to become familiar with the process to establish the correct declination. Some compasses will have a moveable or rotating declination setting or a set screw. Other compasses do not have any declination setting.

Using a Compass to Assist with Map Reading

There may be occasions that a firefighter must determine north on a map. The process is identified below.

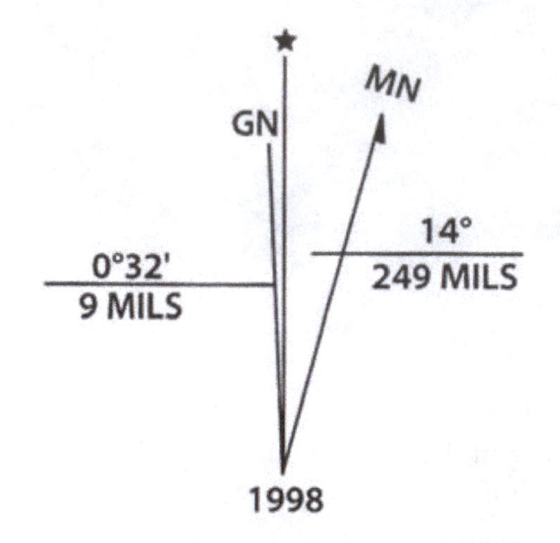

United States Geology Survey

Step 1: Adjust the declination of your compass. Locate on bottom of map or use other references to determine declination.

Courtesy of Brian Henington

Step 2: Rotate the rotating dial (magnetic needle) until the north lines up with the orienting arrow.

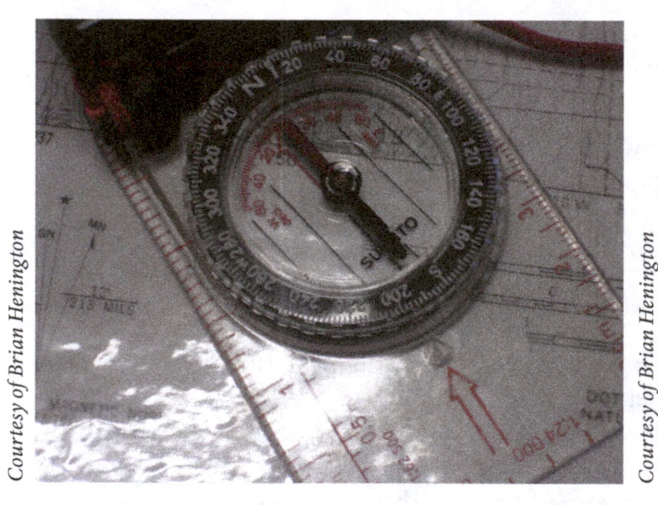

Courtesy of Brian Henington

Step 3: Lay base plate of compass directly on the map. Point orienting arrow to the top of the map.

Courtesy of Brian Henington

Step 4: Rotate the map and compass together until the magnetic needle lines up with the index line. The map will be adjusted toward north.

Azimuth or Bearings

A compass is important in determining an azimuth or bearing. An azimuth or bearing can be explained as "horizontal angles fixing a direction in respect to north, and are expressed in degrees."[4] An azimuth/bearing is based on cardinal directions. An azimuth reading of 0° or 360° is considered north. A 180°

[4] National Wildfire Coordinating Group, *Firefighter Training*, 16.16.

reading is considered south. 90° would be due east and 270° is due west. The bearing reading is primarily used to determine routes of travel to or from a destination. Firefighters are expected to understand how to properly use a compass to determine bearing readings as they navigate to or from a location.

The basic process for determining an azimuth is described in the steps below.

Step 1: Hold the compass away from your body at about shoulder height. Be cautious of any metal on your body or equipment. Metal objects may impact the effectiveness of a compass. (For this example, we will use a 90° bearing.)

Step 2: While you are holding the compass, rotate the dial until 90° is in line with the orienting arrow.

Step 3: With the compass in hand, turn your body slowly until the red magnetic needle lines up with the orienting arrow. This will align you with the 90° bearing.

Back Azimuth

Determining a back azimuth can be done to reverse travel or exit an area using the initial route of travel. This can be done be firefighters to navigate back to their vehicle at the end of their shift. To determine the back azimuth, you must either add or subtract 180° from your original azimuth/bearing reading. You will add 180° to your reading if the original reading is less than 180°. The example we used above would require adding 90° + 180° to give a back azimuth of 270°. If the original bearing is more than 180°, then you will subtract 180°. An example, your original azimuth was 225°. 225° — 180° = 45° back azimuth/bearing.

Using an Azimuth to Determine Bearings on a Map

There may be times that you have to determine the azimuth on a map. This can assist firefighters with travel routes or locating a specific area. The process is illustrated below.

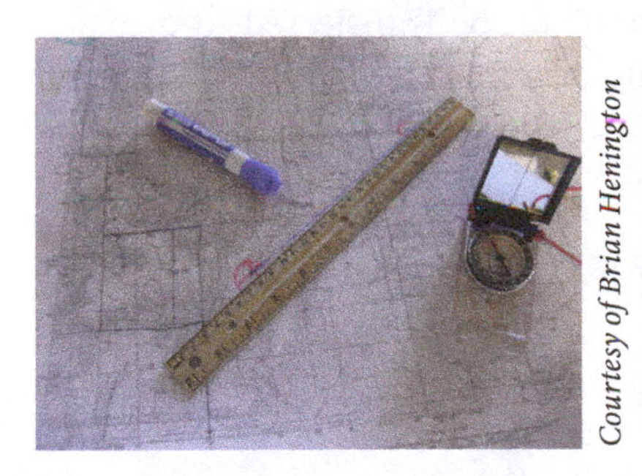

Courtesy of Brian Henington

Step 1: Determine and orient the map to true north. Use the compass edge or other measurement tool to identify your starting point and your destination.

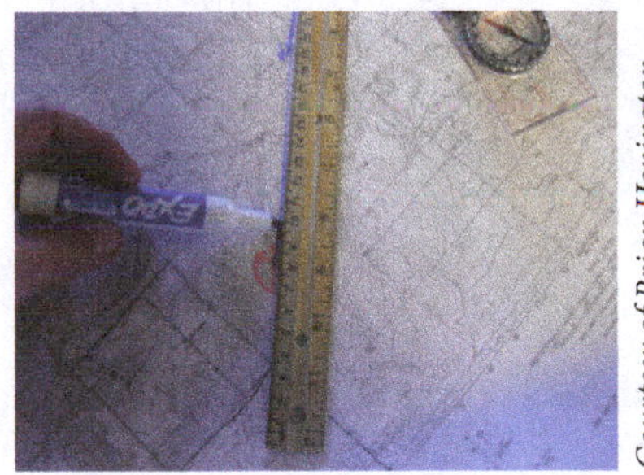

Courtesy of Brian Henington

Step 2: Draw a line between the starting and finish point.

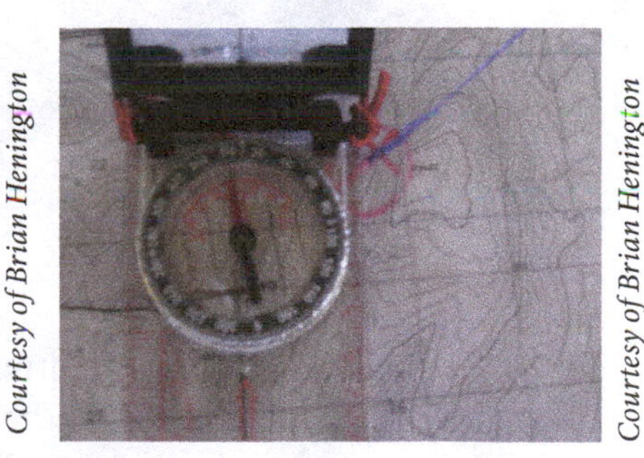

Courtesy of Brian Henington

Step 3: Place the edge of the compass along your drawn line. Turn the compass dial until the orienting line points north.

[4] National Wildfire Coordinating Group, *Firefighter Training*, 16.16.

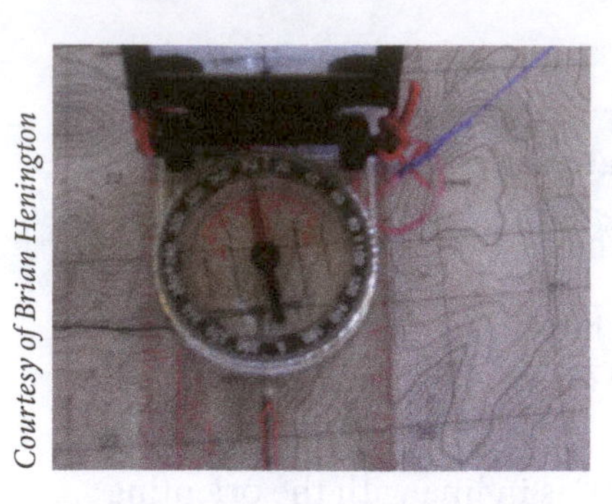

Courtesy of Brian Henington

Step 4: Determine the azimuth reading in relation to true north. Orient your compass to the azimuth reading you determined.

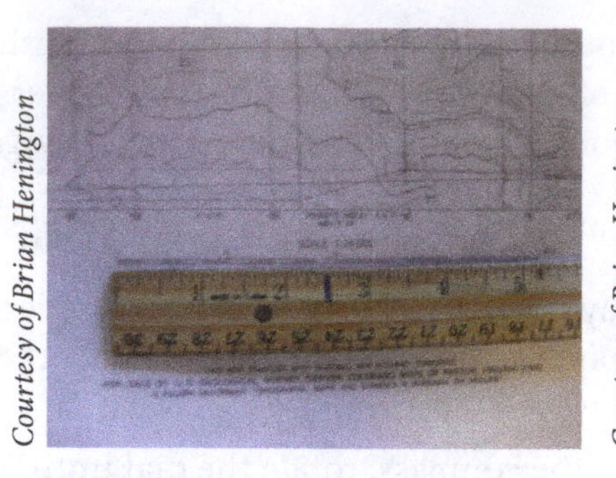

Courtesy of Brian Henington

Step 5: You can also determine the distance required to your location by using the scale of your compass and the scale of the map.

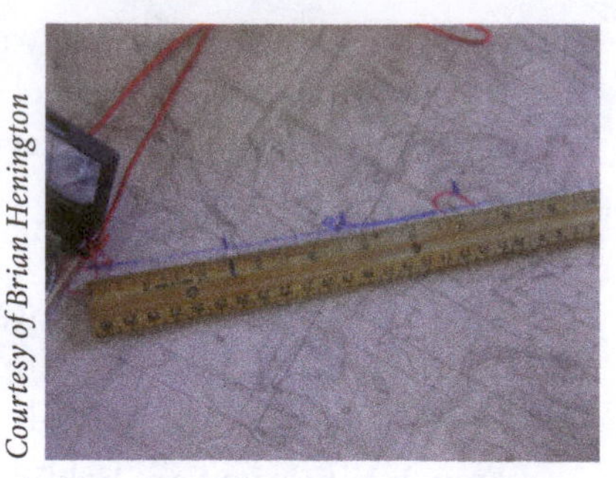

Courtesy of Brian Henington

Note: the distance between the two locations will not consider topography or terrain features. The distance would be as the "crow flies."

GLOBAL POSITIONING SYSTEM (GPS)

© Sander van der Werf/Shutterstock.com

The global position unit (GPS) is a useful tool for firefighters. They range in price and capabilities and are available from numerous manufacturers. If you are assigned a GPS unit, become familiar with the user guide before you try to use the system on the fireline. GPS units are used by firefighters to determine exact locations, fire size, fire line length, and location, and can download information to be used with GIS. This information can be transferred to dispatch centers, other firefighters, or transferred to aircraft (which primarily use latitude/longitude for location determination).

The System

GPS units use latitude and longitude as their primary navigation process. GPS units rely on satellites to determine the exact location of the user on the ground. It is important to note that areas of dense canopies can impact the effectiveness of GPS units.

To better understand how a GPS system determines your location, the land navigation system of latitude and longitude is explained below.

- Latitude/longitude is a global system that precisely identifies locations using the equator as a reference point for latitude and the prime meridian as a reference point for longitude.
- The system is divided or measured in degrees, minutes, and seconds.

- Latitude lines are imaginary lines that run east and west around the earth. The lines are parallel with the equator, which is identified as 0°. Latitude lines are identified by degrees. They are numbered by degrees from their relationship (north and south) from the equator. The latitude lines are further broken down from degrees into minutes and seconds. An example of latitude would be: 34° Degrees, 26 minutes, 34 seconds North.
- Longitude Lines are imaginary lines that run north and south through the north and south poles. These lines are essential to identify degrees from east and west. The dividing line between east and west is called the prime meridian (located in Greenwich, England). Longitude lines are not parallel and do not maintain any set value distance. Longitude lines are also identified in degrees, seconds, and minutes. An example of longitude would be: 108° Degrees, 34 minutes, 52 seconds West.
- Describing latitudes/longitudes: The information below provides the proper way of reporting latitude and longitude readings.

Most common:	32° 71' 22" North (latitude)
	103° 14' 06" West (longitude)
Degrees/Decimal/Minutes:	32° 71.22' North
	103° 14.06' West
Decimal Degrees:	32.7122° North
	103.1406° West

ICS MAPPING SYMBOLS

It is important for firefighters to understand the standard symbols used in the Incident Command System (ICS). Understanding these symbols will ensure you are properly reading a map and identifying key features and components. The table below depicts some of the standard ICS symbols used on maps. For more information, refer to the *Wildland Fire Incident Management Field Guide.*

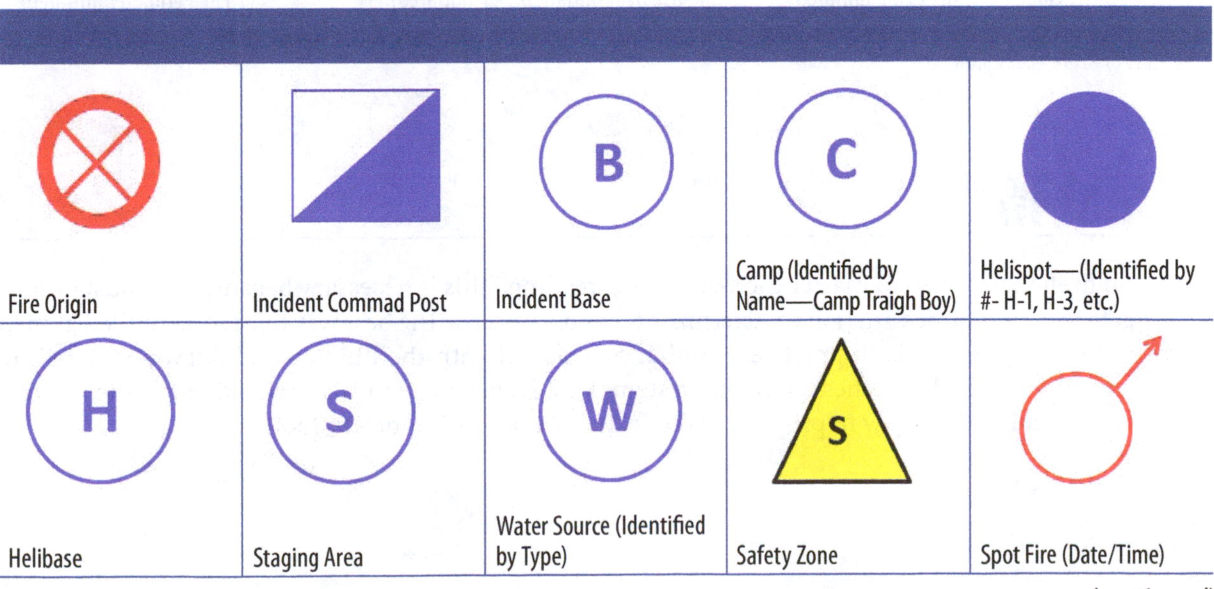

(continued)

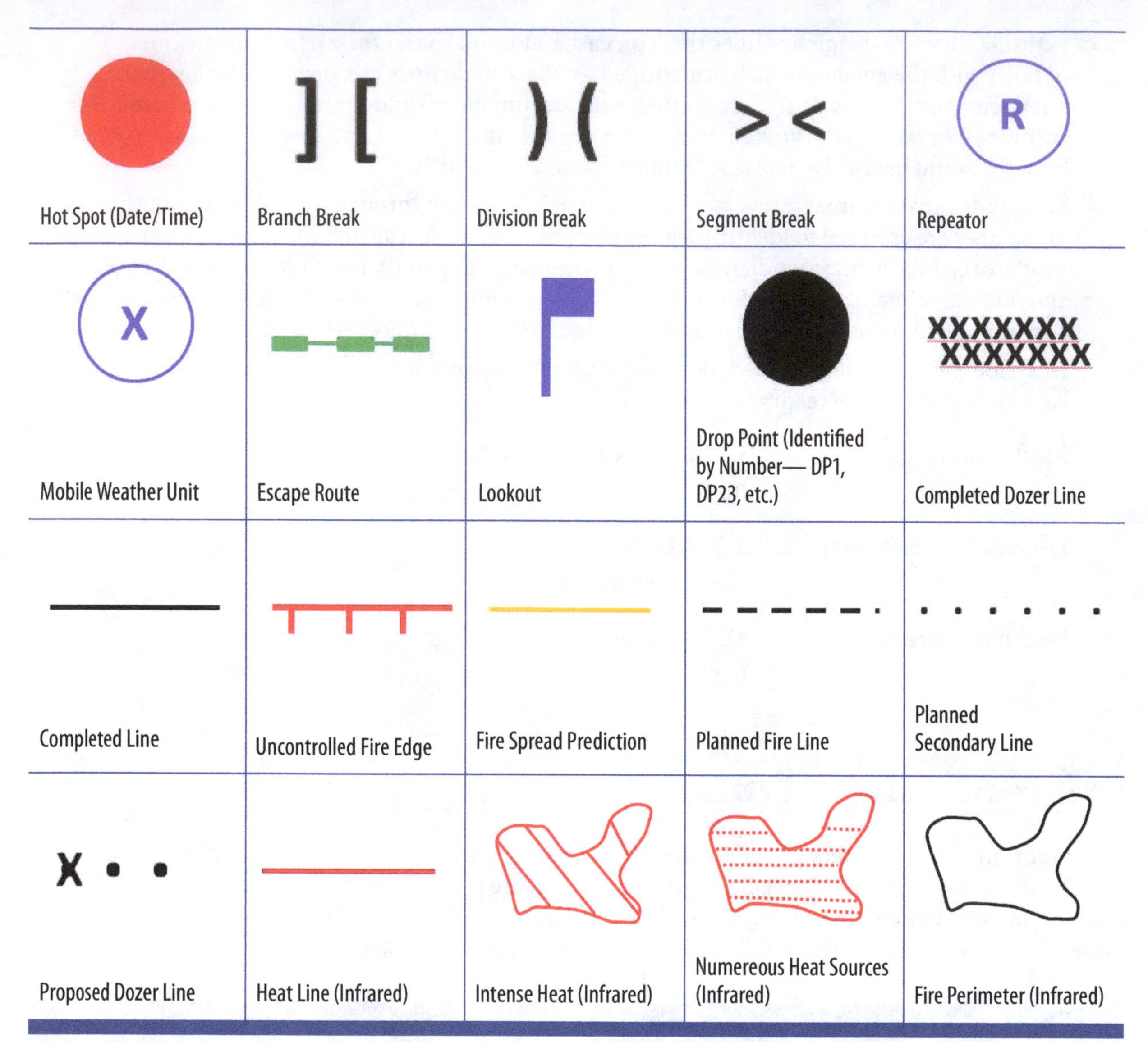

Courtesy of Brian Henington

SUMMARY

Wildland firefighters need to master the skill of map reading. This is necessary because you must be able to navigate to a specific destination or determine areas of concern. The best way to become skilled at map reading and compass use is to practice. Familiarize yourself with the Public Land Survey System and practice calling in fire dispatches using this system. Finally, become familiar and skilled at the use of a compass and understand how to properly and effectively use this important tool.

KNOWLEDGE ASSESSMENT

1. Identify five types of maps used on wildland fires.
2. Define a contour line.
3. What do contour lines that make V's indicate?
4. What do contour lines that make U's indicate?
5. What land descriptive system is most used by fire managers?
6. The PLSS is based on three concepts. What are they?
7. A township is how many square miles?
8. A section is how many square miles and how many acres?
9. An acre is about the size of a ______________. An acre consists of ______________ square feet.
10. True/False: A compass provides a north reading in relation to true north.
11. GPS units identify a location based on ______________.

EXERCISES

1. Complete the following exercise using the PLSS system of legal land description. A mom and dad own one section of land. They have four sons. They want to give each son the equal amount of property; however, they want to keep a quarter section for themselves. Illustrate how you would equally divide the property for each son and the mom and dad. Do not provide a math solution. Draw the solution.

BIBLIOGRAPHY

National Wildfire Coordinating Group. *Firefighter Training, S-130*. Boise: National Wildfire Coordinating Group, 2003.

United States Geological Survey. *The Public Land Survey System (PLSS)*. Washington, D.C.: United States Department of the Interior. http://nationalmap.gov/small_scale/a_plss.html#one.

CHAPTER 18

THE NEXT STEP

OVERVIEW

Courtesy of Brian Henington

The field of wildland fire management is large, complex, and dynamic. In this textbook you have been introduced to many concepts and ideologies that are involved in fire suppression. If this is your first exposure to wildland fire, it is not uncommon to feel overwhelmed. Remember, the intent of this book is to introduce you to the basic concepts that an entry-level firefighter should know and be familiar with before fighting fires. As you continue your firefighting career, the concepts you have learned in this book will be elaborated and expanded on in more and more detail. To be an effective entry-level firefighter, you must understand the basic concepts that influence fire behavior and how those influence(s) can impact firefighter safety. As a firefighter, you should prioritize the following concepts as the most important to your safety and general welfare.

1. **Rules of Engagement:** Understand how these proven and established rules can and will protect you from harm.
2. **LCES:** Understand these four concepts and ensure they are used at all times.
3. **Fire Weather:** Understand the impacts weather events and wind have on fire behavior. Create a routine where you check weather conditions before each shift. Do not rely on somebody else to advise you on what the weather will be today; find out for yourself.
4. **Topography Influences:** Understand how certain terrain features will influence fire behavior. Avoid box canyons and other dangerous terrain features.
5. **Hazard Identification:** Become familiar with the hazards in your area. If you are from New Mexico and are dispatched to a fire in Oregon, take time to research the unique hazards of the new area. Finally, avoid subjecting yourself to objective hazards.
6. **Communicate:** Communicate hazards to other firefighters. Use flagging, notification systems, or other communication measures to notify other firefighters of the hazard you have encountered. Do not assume other firefighters will see what you are seeing.
7. **Student of Fire:** As Paul Gleason said, “Become a student of fire.” Do your part to continue learning about this dynamic and dangerous profession. Do all you can to make yourself a more informed, educated, and competent firefighter.
8. **Continue Your Training:** Seek additional fire training. Purse a degree in fire science or a degree in fire ecology. Anything you can do to make yourself more educated about fire and fire behavior will improve your ability to be an effective firefighter.

WHAT IS THE NEXT STEP?

The section below has some recommendations for the next step in wildland firefighting. Jobs in wildland firefighting are becoming more and more competitive. Enhance your skill base and pursue specialized training to help you become more marketable and a competitive candidate.

1. Apply for jobs.

   Experience is the most important component you can have as a wildland firefighter. All the classroom training in the world will not provide as much improvement as real-life, live fire activities. Get your application ready. Update your resume and begin to create a portfolio on yourself.

2. Volunteer with a fire department.

   If you are not able to get a job, volunteer with a fire department. There are thousands of volunteer fire departments across the United States that are always looking for quality individuals. The beauty of this system is that most volunteer fire departments will pay for you to attend training as well as providing you with real life experience.

3. Pursue a degree in fire science.

   Improve your knowledge base. A college graduate has to learn important organization and self-discipline skills to graduate. A completion of a degree tells your future fire employer that you have learned the soft skills necessary to be an effective wildland firefighter. In addition, the recent trend of fire agencies is to require a degree in order to be considered for promotion. Expect this trend to continue as fire agencies are recruiting a more professional and educated workforce.

4. Gain a specialized skill.

   Learn or become certified in a specialized skill to make you more marketable. Some suggestions include:

   a. Medical Certification: Pursue an EMT with basic or higher medical certification.
   b. Small Machine Mechanics: Learn how to work on small machines like portable pumps, chainsaws, weed trimmers, etc.
   c. Welding: Great asset to have.
   d. Auto mechanics: Vehicles must be maintained or repaired.

5. Participate in National Wildfire Coordinating Group and FEMA trainings.

   Become a certified wildland firefighter. This textbook coincides with NWCG S-190, S-130, L-180, and FEMA I-100. If you have not taken these four classes through a certified entity, then take them as soon as you can.

Incident Command System

I-100.B Introduction to Incident Command System	Note—Self Study
IS-700.A National Incident Management System	Note—Self Study
I-200.B ICS for Single Resources and Initial Action Incidents	Note—Self Study and/or Classroom

Suppression Related

S-130	Firefighter Training
S-133	Look Up, Look Down, Look Around
S-134	LCES: Lookouts, Communications, Escape Routes & Safety Zones
S-190	Introduction to Wildland Fire Behavior
L-180	Human Factors on the Fireline

Specialized Skills—Fire Related

S-211	Portable Pumps and Water Use
S-212	Wildland Fire Chain Saws

SUMMARY

An effective and safe wildland firefighter prioritizes learning and developing his or her personal skill base. Consider the elements covered in this chapter as recommendations to help you pursue a career in the field of wildland firefighting.

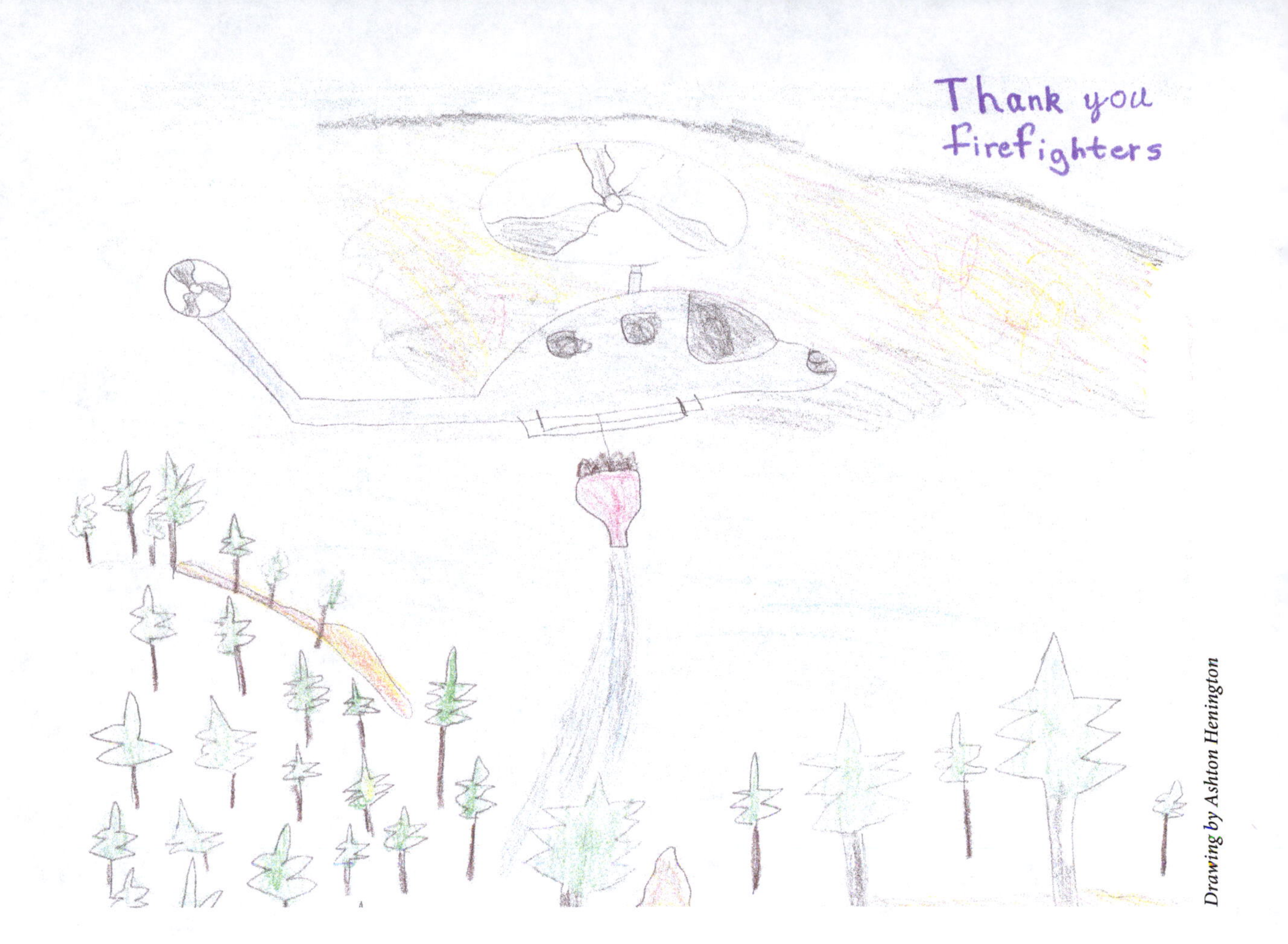

Drawing by Ashton Henington

Appendix A

RECOMMENDED GEAR FOR A WILDLAND FIRE ASSIGNMENT (CHECKLIST)

Fire Pack or Fireline Related

- ____ *Incident Response Pocket Guide*
- ____ Fire shelter
- ____ Fire shirt
- ____ Leather gloves (extra set)
- ____ Hearing protection
- ____ Toilet paper
- ____ Water (canteens and/or camelback)
- ____ Hard hat shroud
- ____ Compass
- ____ Sunscreen lotion
- ____ Matches or Lighter
- ____ Hand sanitizer
- ____ First Aid Kit
- ____ Hard hat with chin strap
- ____ Eye protection
- ____ Fire pants
- ____ Bandanas
- ____ Food
- ____ Flat file
- ____ Signal mirror
- ____ Batteries (AA and AAA)
- ____ Lip balm
- ____ Extra boot laces
- ____ Headlamp
- ____ Space blanket

Other Items:

- ____ GPS unit
- ____ Radio chest pack
- ____ Chainsaw chaps
- ____ Tools for working on specialized equipment
- ____ Hand-held radio
- ____ Spanner wrench
- ____ NFPA-approved fire jacket

Clothes/Gear

- _____ Fire pants (minimum of 2)
- _____ Multi-tool
- _____ Pocket knife with serrated edge
- _____ Belt (leather recommended)
- _____ Undershirts (tee-shirts) (5–8)
- _____ Hooded sweatshirt
- _____ Baseball cap
- _____ Wristwatch
- _____ Driver's license
- _____ Defensive driving certificate
- _____ Red card
- _____ Blue pens
- _____ Fire shirt (minimum of 2)
- _____ Carabiner (several)
- _____ Socks (10–14 pair)
- _____ Underwear (10–14)
- _____ Jacket
- _____ Beanie (stocking hat)
- _____ Jeans (1)
- _____ PT clothes
- _____ Camera
- _____ Book (for down time)
- _____ Watch
- _____ Flashlight

Personal Items (Including Sleeping Items)

- _____ Tent
- _____ Sleeping pad
- _____ Alarm clock
- _____ Shorts
- _____ Tennis shoes
- _____ Toothbrush
- _____ Toothpaste
- _____ Shampoo
- _____ Deodorant
- _____ Baby wipes
- _____ Nasal decongestant
- _____ Soap
- _____ Moleskin
- _____ Athletic tape
- _____ Credit/debit card
- _____ Cell phone with portable charger
- _____ Sleeping bag with storage bag
- _____ Towel
- _____ Flip Flops (showering)
- _____ Small pillow
- _____ Extra prescription glasses
- _____ Razors
- _____ Sunscreen (extra)
- _____ Lip balm (extra)
- _____ Comb/Brush
- _____ Tweezers
- _____ Anti-chaffing cream or powder
- _____ Body glide
- _____ Blister cream
- _____ Cash (at least $50.00)
- _____ Personal hygiene items for women
- _____ Prescription medicines (enough for 14 days)

CHAPTER 1

Knowledge Assessment

1. Explain the difference between a wildfire and a wildland fire.

   Answer: **Wildfire:** An unplanned, unwanted wildland fire including unauthorized human-caused fires, escaped wildland fire use events, escaped prescribed fire projects, and all other wildland fires where the objective is to put the fire out.

   Wildland Fire: Any non-structure fire that occurs in vegetation or natural fuels. Wildland fire includes prescribed fire and wildfire.

2. Define wildland fire management.

   Answer: The activities taken by wildland fire managers to suppress, manage, and/or control wildland fires.

   - This concept also includes the actions taken by fire managers to direct and guide individuals participating in fire suppression activities.
   - Wildland fire management may also include the activities taken to reduce the potential for catastrophic fire (e.g., fuel mitigation, thinning, prescribed fire, etc.) or activities to improve forest health and ecosystem management.

3. What tool do wildland firefighters use to document work activities, competencies, and experience?

   Answer: Position Task Book.

4. What is the title of an entry-level firefighter?

   Answer: Firefighter Type 2 (FFT2).

5. Identify three references used to support wildland fire activities.

   Answer:

   - *Incident Response Pocket Guide*
   - *Wildland Fire Incident Management Field Guide*
   - National Incident Management System: Wildland Fire Qualification System Guide—PMS 310-1

6. Out of the three reference materials, which one is considered the most important?

   Answer: *Incident Response Pocket Guide.*

7. What wildland fire organization develops and maintains training standards and curriculum?

   Answer: National Wildfire Coordinating Group.

8. What is the primary function of the National Interagency Fire Center?

   Answer: The purpose of this entity is to provide coordination among wildland fire agencies and reduce the duplication of specific efforts and activities.

9. What role does the National Weather Service play in wildland fire management?

   Answer: The NWS provides critical fire weather information, predictions, forecasts, and outlooks.

10. What are the three core values of the Wildland Fire Leadership Program?

    Answer: Duty, respect and integrity.

11. What is *Six Minutes for Safety*?

    Answer: *Six Minutes for Safety* is a safety program that provides daily safety briefings related to fire operations and safety concepts.

Exercises

1. Research the state agency that is responsible for wildland fire suppression in your state. How is the agency established (e.g., state forestry or natural resource department)? How large is it? Provide a web link of the agency for your instructor.

   Answer: Student should identify their state's wildland fire organization and identify what the agency is referred to. The student should also identify how large the organization is. A link to the web page should be provided for the instructor.

2. Research the Wildland Firefighter Foundation using the Internet. What function of the organization really stood out to you?

   Answer: Analytical answer. The intent of this question is to familiarize the student with this organization.

CHAPTER 2

Knowledge Assessment

1. What section of the ICS conducts tactical operations to carry out the plan and directs all tactical resources?

   Answer: Operations.

2. True or False: The Incident Commander is always the most qualified individual on a wildland fire.

   Answer: False. Not always. Type 5 incidents can be managed by a Type 5 Incident Commander even though a qualified Type 3 Incident Commander is on scene.

3. On a wildland fire, the Medical Unit is in what section?

   Answer: Logistics.

4. True or False: Wildland firefighters, structural firefighters, and law enforcement officers can be organized under the same incident command system.

   Answer: True.

5. True or False: Divisions and Groups are at the same organizational level and can work together on an incident.

   Answer: True.

6. True or False: The Incident Command System is not designed for small incidents.

   Answer: False.

7. True or False: Deputies must always be as qualified as the person for whom they work.

   Answer: True.

8. In what section and unit can you locate hand tools, sleeping bags, safety gear, and 1 ½" hose?

   Answer: Logistics Section/Supply Unit.

9. In ICS, communication is in ________________.

   Answer: clear text.

10. The five major management activities (positions) around which the ICS is organized.

    Answer:

    - Command
    - Operations
    - Planning
    - Logistics
    - Finance/Administration

11. Span of control may vary from ________________ to ________________, with a reporting element of ________________ subordinates to ________________ supervisor.

    Answer: 3 to 7; 5 to 1.

EXERCISES

1. The General Staff consists of:

   Answer:

   - Operations Section Chief
   - Planning Section Chief
   - Logistics Section Chief
   - Finance/Administration Section Chief

2. What are the three (3) major activities (positions) of the command staff:

   Answer:

   - Safety Officer
   - Liaison Officer
   - Information Officer

3. Division Supervisors have __________________ responsibility.

   Answer: geographic

4. Group Supervisors have __________________ responsibility.

   Answer: functional

5. Identify the common ICS responsibilities of all emergency responders.

   Answer:

   - Receive an assignment.
   - What job are you being dispatched to perform?
   - Officially check in.
   - Use clear text on the radio.
   - Briefings: Ensure you obtain a briefing from your direct supervisor. If you are a supervisor, ensure you brief the people working for you.
   - Organization: Ensure you are organized and have the appropriate tool(s) for the specific task.
   - Work with the adjoining resources and brief your replacement.
   - Complete all required forms.
   - Demobilize according to the plan.

CHAPTER 3

Knowledge Assessment

1. What is the definition of fire?

Answer: Fire is a set of chemical reactions that produce heat and light and an occurrence in which something burns: the destruction of something (such as a building or a forest) by fire.

2. What time of the day is considered the burn period?

Answer: Between 2 pm and 6 pm.

3. What are the three elements of the fire triangle?

Answer: Oxygen, heat and fuel.

4. Out of the three elements of the fire triangle, which one can be controlled or mitigated the most by wildland firefighters?

Answer: Fuel.

5. Name and briefly define the nine parts of a wildland fire. Which part is considered the most dangerous? Which part is where the point of origin typically occurs?

Answer:

1. Head of a Fire: The head of the fire is the side of the fire that is moving the fastest.
2. Point of Origin: The exact location where ignition occurred and sustained combustion.
3. Rear of a Fire: The opposite of the fire head.
4. Flank of a Fire: The flanks of a fire are included in the fire perimeter and typically run parallel to the fire spread.
5. Fire Perimeter: The complete outside edge of a fire.
6. Fingers: A finger of a fire is defined as the long narrow extensions of a fire projecting from the main body.
7. Pockets: Pockets are the area(s) between fingers.
8. Islands: Pockets occur within the main body of the fire.
9. Spot Fire: Spot fires occur outside or away from the main fire.

A. Which part is considered the most dangerous?

Answer: Head of the Fire.

B. Which part is where the point of origin typically occurs?

Answer: Rear of the Fire.

6. Which fire behavior term is considered the most dangerous and presents the most complexity to fire suppression activities?

Answer: Crown Fire.

7. Define the term contained and controlled.

Answer:

- Contained or containment of a fire identifies the progress firefighters have made in their efforts to suppress a fire.
- Controlled is the completion of control line around a fire.

8. How many chains are in a mile? How many chains would be in 2.5 miles?

Answer: 80 chains in a mile; 200 chains in 2.5 miles.

9. List the three components of the heat transfer process. Which two are we most concerned with as it relates to fire suppression activities?

Answer: Convection, radiation, and conduction.

We are most concerned with convection and radiation.

EXERCISES

1. On a piece of paper, draw and identify the nine parts of a fire. In addition to the parts of the fire include at least two anchor points (identify them by a large A).

Answer: The student shall draw and identify the nine parts of a fire and include two anchor points.

2. Try an experiment. Take a candle (ensure you are somewhere safe and the candle flame will not touch available fuel). Light the candle. Observe the convective heat lifting vertically above the candle. Now feel for heat next to the candle (at the sides).This would be radiant heat. Which heat (above or to the side) is more intense?

Answer: The instructor should determine completion of the exercise by having the students explain the experiment in class.

CHAPTER 4

Knowledge Assessment

1. Which aspect presents the greatest opportunity for fire spread? Which aspect (under normal conditions) would have the least opportunity for fire behavior?

   Answer:

   - South, southwest or southeast.
   - North.

2. Why do fires travel up slopes faster than they do down slope?

   Answer: Radiant and convective heat travel up the slope, which causes preheating and drying of fuels above the flaming front. Flames are closer to unburned fuel above them, causing fire spread to occur very rapidly.

3. Why are box canyons dangerous to firefighters?

   Answer: A box canyon is dangerous because it funnels wind and fire behavior, creating a chimney effect.

4. Horizontal continuity plays an important role in fire spread. Name the two categories. Out of the two categories, which one would have higher fire spread?

   Answer:

   - Uniform or continuous fuels are considered fuels that have direct or immediate contact with adjoining fuels.
   - Patchy fuels do not have immediate contact with the adjoining fuels.
   - Uniform or continuous fuels would have higher fire spread.

5. List the three categories of vertical arrangement.

   Answer:

   1. Ground fuels
   2. Surface fuels
   3. Aerial fuels

6. Explain how ladder fuels allow a fire to travel from the surface to the crowns of trees or brush.

   Answer: They allow fire to travel from the surface to the tops or canopies of trees.

7. What is temperature and why is it important for firefighters to monitor?

   Answer: Temperature provides a measurement of hot and cold. Temperature will vary depending on aspect, time of the day, and/or elevation changes.

8. What time of day would you expect temperature to be at the highest? When would relative humidity be at its lowest?

   Answer: Midafternoon or the burn period for both.

9. What is the relationship between temperature and relative humidity?

Answer: Indirect or inverse relationship

10. What event or occurrence is usually associated with stable atmospheric conditions?

Answer: Inversion.

11. How do you determine the direction of wind?

Answer: Wind direction is determined by the direction the wind is blowing from.

12. If a fire was experiencing southwest winds, what direction would the head of your fire be traveling?

Answer: Northeast.

13. List four local winds.

Answer:

- Upslope winds
- Downslope winds
- Up-valley winds
- Down-valley winds

Exercises

1. Research dangerous wildland fuel types in your geographical area. Why are they dangerous? Have they been involved in firefighter fatalities in your area?

Answer: Should be analytical and should provide examples of why the fuels are dangerous.

CHAPTER 5

Knowledge Assessment

1. What are the four Red Flag Events that can lead to a Red Flag Warning or Watch?

   Answer:

   1. Dry lightning
   2. Wind
   3. Humidity
   4. Low Fuel Moisture

2. What agency is responsible for issuing Red Flag Warnings or Watches?

   Answer: National Weather Service.

3. What is a cold front? What are the dangers associated with cold fronts? What major fire fatality occurred during an approaching cold front?

   Answer:

   - A cold front is defined as the transition zone where a cold air mass is replacing a warmer air mass.
   - South Canyon Fire.

4. Thunderstorms are very dangerous to firefighters. What are the two major concerns identified in this chapter?

   Answer:

   1. Storms will produce lightning
   2. Downdraft winds

5. What historical firefighter fatality occurred due to an approaching thunderstorm?

   Answer: Yarnell Hill Fire.

6. List the common types of foehn winds that occur in the United States.

   Answer:

   - Santa Ana and Sundowner
   - Chinook wind
   - Wasatch wind
   - Mono or North
   - East wind

7. Why are firewhirls dangerous?

   Answer: Firewhirls are dangerous because they can move in any direction and they are a major source for transporting fire brands which will create spot fires.

8. List the seven Look Up, Look Down, Look Around factors that must be monitored by wildland firefighters.

   Answer:

   1. Fuel Characteristics
   2. Fuel Moisture
   3. Fuel Temperature
   4. Topography
   5. Wind
   6. Atmospheric Stability
   7. Fire Behavior

EXERCISES

1. Using your *Incident Response Pocket Guide (IRPG)*, find thunderstorm safety. What page of the IRPG did you locate this safety checklist? List for your instructor the steps involved in this checklist.

   Answer: 2014 Version on *IRPG*, pg. 21. Students should identify the steps in writing.

2. Using the National Weather Service as a resource: Locate a lightning related fatality that occurred in your area. How did it occur? What activity was the individual(s) doing when they were struck?

   Answer: Students should document their findings and summarize the event in writing.

3. Research the National Fire Center's *Historical Wildland Firefighter Fatality Reports*. Search the database by type of accident. How many firefighters died in the line of duty due to lightning strikes? What state(s) did they occur in?

   Answer: Students should provide up-to-date data related to this type of fatality. Students should also identify the states where the fatalities occurred.

4. Conduct an Internet search on firewhirls. Provide your instructor with the website. You do not have to write a discussion on them. The question we have for you is: What did you think about this phenomenon?

   Answer: Answers may vary. Use instructor discretion.

CHAPTER 6

Knowledge Assessment

1. List the standard personal protective equipment required to participate in wildland fire activities.

Answer:

- Hard hat
- Flame-resistant clothing
- Eye protection
- Hearing protection
- Gloves
- Fire boots
- Fire shelter
- Canteen or camelback
- Chinstrap
- Fire pack or harness to carry fire shelter
- Headlamp
- First aid kit
- *Incident Response Pocket Guide*

2. Explain the importance of accountability and maintenance of personal protective equipment.

Answer: The equipment or PPE belongs to the agency. We must prioritize maintaining the equipment and PPE at the highest standard to ensure the most effective use. Fire equipment and PPE are very expensive.

3. What are the minimum standards for a wildland firefighter's boot?

Answer:

- 100% leather
- Minimum of 8" Tall
- No steel toe
- Sole: Deep tread (traction)
- Leather shoe laces

4. What are the national standard weight limitations for a firefighter's gear?

Answer:

- Total weight—65 lbs.
- Overnight (personal gear—45 lbs.
- Fire pack/web gear—20 lbs. (excluding water)

5. Name four of the seven firefighter responsibilities.

Answer: Any four of the following:

- Performs manual and semi-skilled labor.
- Ensures the objectives and instructions are understood.
- Performs work in a safe manner.
- Maintains self in the physical condition required to perform the duties of fire suppression.
- Keeps personal clothing and equipment in serviceable condition.
- Reports close calls, accidents, or injuries to supervisor.
- Reports hazardous conditions to supervisor.

6. Why is it important to maintain professionalism?

Answer: A professional firefighter is one who acts in a manner that others see as honorable. Professional firefighters should be trustworthy, honest, respectful, and accountable for their own actions. The public holds firefighters in high regard.

7. Explain cultural differences and how they impact wildland fire activities.

Answer: Firefighters are from different backgrounds, genders, races, ethnicities, religious beliefs, and/or political affiliation. Firefighters should respect the fact that people are entitled to their own beliefs. When someone is offended because of their cultural differences, they can cause distractions and conflicts.

8. What are three steps involved with successful teams?

Answer: Any three of the following:

- Know yourself and seek improvement.
- Be technically and tactically proficient.
- Comply with orders and initiate appropriate actions in the absence of orders.
- Develop a sense of responsibility and take responsibility for your actions.
- Make sound and timely decisions and recommendations.
- Set the example for others.
- Be familiar with your leaders their jobs, and anticipate their requirements.
- Keep your leaders informed.
- Understand the task and ethically accomplish it.
- Be a team member—but not a "Yes Person."

9. What are three activities common with unsuccessful teams?

Answer: Any three of the following:

- Bad attitude is obvious among individuals or crew members.
- Laziness. Crew members not pulling their weight and performing their specific job duties to the best of their ability.
- Crews or crew members acting in an arrogant or with a "know it all" attitude.
- Treating other crews or firefighters in a disrespectful manner.
- Cliques among crew members.

- Unethical activities.
- Not taking responsibility for individual's mistakes or failures.
- Passing the mistakes made to other crew members or crews.

10. What does the Pack Test consist of?

Answer: Three-mile hike with a 45 lb. pack, completed under 45 minutes.

11. Is fitness important to the wildland firefighter?

Answer: Yes

12. What are the three heat related injuries/illnesses that can effect firefighters?

Answer:

1. Heat Cramps
2. Heat Cramps
3. Heat Stroke

EXERCISES

1. Research, develop and plan for a rigorous workout or fitness program. Start immediately so you can ensure you are in the best possible shape to handle the rigors of this job.

Answer: No answer is required for this question unless the instructor wants a fitness plan from the student. This question may best be asked during group discussions.

2. Using your *Incident Reponses Pocket Guide*, locate and familiarize yourself with heat-related injuries. Have you experienced any of the systems or injuries before? Select one type of heat-related injury and list for your instructor the symptoms and treatments and research one incident that involved a heat-related injury. (Provide reference source and type of injury; it does not have to involve a firefighter).

Answer: Response should include at least one heat-related symptom and treatment. Students should also provide a summary of an event that involved a heat-related illness.

3. Before you get your first fire assignment, we want you to consider these situations that impact wildland firefighters. The purpose of this exercise is to have a plan in place so you can implement it once you get a fire assignment. 1) Have you made arrangements for somebody to pay your bills, water your lawn, etc.? 2) Do you have arrangements for your pets? 3) Have you told your love ones that you love them?

Answer: This question does not have a specific answer. It is intended to have students think about the preparations needed before taking a fire assignment.

CHAPTER 7

Knowledge Assessment

1. What is a resource type and kind? Provide an example of each.

   Answer:

   - Resource Kind: What the resource is.
   - Resource Type: Describes the size, capability, and staffing qualifications of a specific kind of resource.

2. What are four types of resources used on a wildland fire?

   Answer: Any four of the following:

   - Fire engines
   - Hand crews
   - Aircraft
   - Heavy equipment
   - Dozers
   - Helitack
   - Firing Crews
   - Felling Crews

3. What is a single resource?

   Answer: A single resource is described as "an individual, a piece of equipment and its personnel complement, or a crew or team of individuals with an identified work supervisor that can be used on an incident."

4. Explain the difference between a task force and a strike team.

   Answer:

   - Strike Team is defined as specified combinations of the same kind and type of resources, with common communications, and a leader.
   - A Task Force is defines as any combination of single resources assembled for a particular need, with common communications and a leader.

5. How many resources are involved in a strike team or task force? What is the ideal ratio of supervisor to subordinate/resource?

   Answer: Two to seven with the ideal ratio of five resources with one assigned supervisor.

6. What is the major difference between Interagency Hotshot Crews and other hand crews?

   Answer: Interagency Hotshot Crews are considered national resources whereas other crews do not have this consideration. They have higher training standards, experience levels, and physical fitness levels.

7. How do smokejumpers access a fire?

   Answer: By parachute.

8. What engine is considered the work horse on wildland fires?

Answer: Type 6.

9. What is the biggest threat or risk related to a helicopter?

Answer: Rotors.

10. How many accidents and fatalities have been associated with driving activities on a fire?

Answer:

- 139 driving accidents
- 110 have resulted in fatalities

11. A driver of fire equipment cannot operate the apparatus for no more than _____ hours on a given day.

Answer: 10 hours

12. Should a driver of fire equipment use a cell phone or radio? If not, who should?

Answer: No. A co-pilot should use the cell or radio.

13. What are some hazards associated with foot travel?

Answer: Any of the following:

- Tree Related
  - Snags
  - Leaning trees
  - Whipping branches
  - Widow-makers
  - Stump holes
- Environmental
  - Darkness
  - River or stream crossings
  - Rolling rocks
  - Rolling logs
  - Uneven terrain
  - Rocks or boulders
  - Snakes
  - Poisonous insects
  - Poisonous plants
- Human or Firefighter Related
  - Sharp hand tools
  - Other firefighters' tools
  - ATVs
  - UTVs
  - Vehicles
  - Heavy equipment
  - Aircraft

14. What is the minimum distance firefighters must maintain between each other when walking to their assigned work area?

Answer: 10 feet.

EXERCISES

1. Research the requirements and job responsibilities of a Smokejumper. You can conduct this research using outside references, such as searching the internet.

Answer: Student should be able to provide the instructor with the responsibilities of a smoke-jumper. Reference should be provided.

2. Explain the qualifications of chainsaw operators. What is the basic level of training required by the National Wildfire Coordinating Group? How does a sawyer move from one category to the next?

Answer:

1. A Faller—Up to 8" DBH
2. B Faller—Up to 24" DBH
3. C Faller—Unlimited—can also be based on hazard rating or DBH specification
   - Basic Training by NWCG—S-212 Wildland Fire Chainsaws
   - Sawyer moves through experience and a certification process

3. Conduct an Internet search of a vehicle accident involving a firefighter (either wildland or structural). How did the accident happen? Where was it located? Did you find any that occurred near or where you live? What would be your recommendation to ensure a similar accident does not occur?

Answer: should include:

- Brief summary of the incident
- How accident happened
- Location
- An analysis of mitigation measures while driving

The intent of this question is to increase the awareness of students as it relates to driving a fire-related vehicle.

CHAPTER 8

Knowledge Assessment

1. Define a safety zone.

   Answer: A safety zone is defined as a preplanned area of sufficient size and suitable location that is expected to protect fire personnel from hazards without using fire shelters.

2. Who should function as a lookout?

   Answer: An experienced, competent, and trusted firefighter.

3. What is the preferred safety zone for firefighters? What would be the determining factor if the safety zone would be considered adequate?

   Answer: The preferred safety zone along the fireline is the black or area already burned by the fire.

4. Identify the elements involved in LCES.

   Answer:

   - Lookouts
   - Communications
   - Escape Routes
   - Safety Zones

5. Who did Paul Gleason suggest should be a lookout on an Interagency Hotshot Crew?

   Answer: Hotshot superintendent

6. What is considered Good Black? What is considered Bad Black?

   Answer:

   - "Good Black" is an area that has burned available fuel and lacks more fuel for consumption, including the overstory or brush (what is considered reburn potential).
   - "Bad Black" is an area the fire has not burned completely or has preheated the overstory in such a way that the trees or brush could sustain fire movement and combustion.

7. What is a deployment site? Why is it different from a safety zone?

   Answer:

   - A deployment site is an area used by firefighters to deploy a fire shelter or an area identified as a suitable location to deploy a fire shelter.
   - The deployment site is specifically tied to a fire shelter and should not be considered a safety zone because it lacks the sufficient size to protect firefighters and equipment.

EXERCISES

1. Using your *Incident Response Pocket Guide (IRPG)*, locate LCES and specifically lookouts. What does the *IRPG* recommend for a lookout?

   Answer:

   - Experienced, competent, trusted.
   - Enough lookouts at good vantage points.
   - Knowledge of crew locations.
   - Knowledge of escape routes and safety locations.
   - Knowledge of trigger points.
   - Has map, weather kit, watch, IAP.

2. Locate the Standard Firefighting Orders in your IRPG. On what page can you locate these rules? You should begin to learn and understand the rules immediately. Begin quizzing yourself. Truly understand their intent. You should approach this as if you are going to teach a new group of firefighters how important these rules are.

   Answer: Located on back cover of the *IRPG.*

3. Research the South Canyon Fire Investigation Report. You can locate this document online by conducting a web search of South Canyon Fire Investigation Report. Locate the Witness Statement completed by Tony Petrilli. Mr. Petrilli lists the violations of the Watch Out Situations that occurred on this fire. How many were in violation? What do you think about the violations?

   Answer: This is an analytical question. Students should answer the question critically and provide and short analysis of what they think occurred.

4. View the video produced by the Wildland Fire Lessons Learned Center "Oh, It's Just a Grass Fire." https://www.youtube.com/watch?v=hl1gNlF0JkY. Provide your instructor with a short narrative of the video.

   The answers can vary but students should at least mention:

   - How dangerous grass fires can be.
   - Attacking the fire from the black.
   - Ensure proper PPE is worn at all times, even on grass fires.

CHAPTER 9

Knowledge Assessment

1. The key to any successful fire operation is not to have ________________.

   Answer: available escape options eliminated or compromised.

2. We should always use the ________________ to ensure our safety is not compromised.

   Answer: Rules of Engagement

3. What are trigger points?

   Answer: Geographic points on the ground or specific points in time where an escalation or alternative of management actions is warranted.

4. Who was the first documented firefighter to use an escape fire to survive an advancing fire? What was the name of the fire and what year did it occur? Where did it occur?

   Answers:

   - Wag Dodge.
   - Mann Gulch Fire—1949
   - Montana or northwestern Montana

5. ________________ may be the ideal tool to use when escaping an advancing fire.

   Answer: Vehicles

6. What direction should your vehicle always be pointed?

   Answer: Toward the escape route

7. Can you use a structure as a temporary refuge to escape an advancing fire? What advantage does this have?

   Answer: Yes, this allows you to receive relief from the initial blast of super-heated gases.

8. Is the fire shelter considered mandatory PPE?

   Answer: Yes.

9. True or False: Fire shelters are not intended for firefighters to take greater risks or not to violate the Rules of Engagement.

   Answer: True

10. What two important features does the New Generation Fire Shelter provide?

    Answers:

    1. Reflect radiant heat
    2. Trap breathable air for the firefighter

11. What is the maximum temperature (in Fahrenheit) a fire shelter is considered effective? Is it designed for direct flame contact?

Answers:

- 500° F
- No, it is not designed for direct flame contact

12. How often should you practice deploying your fire shelter?

Answer: All the time. As much as possible.

EXERCISES

1. What concepts can be used when escaping a fire with or without the use of your fire shelter?
    1. Try all options to escape the advancing fire. This may include outrunning the fire, using vehicles, helicopters, or taking refuge in a lake or large river.
    2. What is your best available option? Use it! Seeking refuge in a lake may be appropriate if the lake is deeper than two feet. Make sure firefighters can swim.
    3. When outrunning or moving away from the fire, drop your gear. Your fire gear can slow your retreat time considerably. If you do so, you must keep your fire shelter, water (pull out your camelback or grab a canteen), radio, gloves, and hand tool.
    4. Look for light fuels, such as grass. Light fuels will not produce as much heat as heavy fuels. The residual burning of light fuels is far less than heavy fuels.
    5. A fire shelter could be used as a heat barrier while you are moving. Extensive practice with this technique is highly recommended.

2. If you are ever caught in underburned fuel (in light fuels) with an advancing fire—turn your vehicle into the advancing flames and drive into the black. Explain why this concept is proven.

Answer: The area behind the flames will be burnt with limited residual burning.

3. Is a safety zone and deployment site the same thing? What is the difference?
    - A deployment site involves the use of a fire shelter.
    - A safety zone should be big enough that you can survive a fire without the use of a fire shelter.

CHAPTER 10

Knowledge Assessment

1. What is the number one cause of wildland firefighter fatalities? How may fatalities have been involved with helicopters? How many fatalities from electrocution?

   Answers:

   - Burnover
   - 56
   - 11

2. What is the definition of safety?

   Answers:

   1. The condition of being safe from undergoing or causing hurt, injury of loss.
   2. Freedom from exposure to danger, exemption from injury, and to protect from injury.

3. What are the five (5) steps involved with a safety culture?

   Answers:

   1. Provide honest sharing of safety information without the fear of reprisal from superiors.
   2. Adopt a non-punitive policy toward errors.
   3. Take action to reduce errors in the system. Walk the talk.
   4. Train fire fighters in error avoidance and detection.
   5. Train fire officers in evaluating situations, reinforcing error avoidance, and managing the safety process.

4. What are the two categories of fireline hazards?

   Answer: Subjective and Objective.

5. Which hazard category is controllable?

   Answer: Subjective

6. Controlling the hazard is called ____________________?

   Answer: Mitigation

7. What communication device can you use to identify hazards?

   Answer: Flagging

8. What is the definition of a snag?

   Answer: A standing dead tree or part of a dead tree from which at least the leaves and smaller branches have fallen.

9. What are three byproducts (chemicals) found in smoke?

Answers:

1. Carbon monoxide
2. Carbon dioxide
3. Hydrogen cyanide

10. Does the firefighter have the right to refuse risk? What is it based on?

Answer: Yes, based on safety concerns.

EXERCISES

1. Study Table 10-1. What hazard type related to fatalities stood out to you? Why?

Answer: Student answer should be specific and the analysis should be analytical in nature.

2. Locate Thunderstorm Safety in the *Incident Response Pocket Guide*. How many steps are involved in the process? In addition, how many firefighters have died in the line of duty due to lightning strikes?

Answers:

- 10 steps in the process
- 3 fatalities

3. Familiarize yourself with the Hazard Tree Safety in the *Incident Response Pocket Guide*. In addition, how many firefighters have died in the line of duty due to tree related activities?

Answer: 44

4. Locate How to Properly Refuse Risk in the *Incident Response Pocket Guide*. What are the four factors involved with refusing risk.

1. There is a violation of safe work practices.
2. Environmental conditions make the work unsafe.
3. They lack the necessary qualifications or experience.
4. Defective equipment is being used.

CHAPTER 11

Knowledge Assessment

1. What is situational awareness?

   Answer: An on-going process of gathering information by observation and by communication with others. This information is integrated to create an individual's perception of a given situation.

2. Identify the five steps involved in situational awareness.

   Answer:

   1. Always functioning: The cycle only works if it is constantly being updated and reevaluated.
   2. Evolve and Adapt: The cycle has to evolve with the events of the day to be effective. A good situational awareness cycle helps us to anticipate changes.
   3. Gather Information: Requires constant information gathering. This can be done through personal observations or the observations of others.
   4. Communicate: Communication has to be maintained and prioritized.
   5. Top Priority: Considered the most important step in the risk management process.

3. Who is the sender? Who is the receiver?

   Answer:

   - The sender is the person who initiated the communication.
   - The receiver is the person for whom the communication is intended.

4. Explain direct communication.

   Answer: Direct communication is described as straightforward and precise. The proper use of direct communication will expedite the communication process and facilitate effective transfer of information.

5. What are the five common responsibilities of communication?

   Answer:

   1. Brief others
   2. Debrief your actions
   3. Communicate hazards to others
   4. Acknowledge messages and understand intent
   5. Ask if you do not know

6. What is the definition of standard operating procedures? Give five examples of SOPs used on wildland fire activities.

Answer: Standard operating procedures are specific instructions clearly spelling out what is expected of an individual every time they perform a given task. A standard operating procedure can be used as a performance standard for tasks that are routinely done in the operational environment.

- Standard Fire Orders
- Helicopter hand signals
- Dispatch protocol
- Dozer hand signals
- Radio call signs
- Radio frequencies

7. What are the four questions involved in an AAR?

Answer:

1. What was planned?
2. What actually happened?
3. Why did it happen?
4. What can we do better next time?

8. The AAR is only effective if all members are ________________.

Answer: Involved

9. What are the five steps in the decision making process?

Answer:

- Step 1: Situational Awareness
- Step 2: Hazard Assessment
- Step 3: Hazard Control
- Step 4: Decision Point
- Step 5: Evaluate

10. What are the hazardous attitudes identified in this chapter?

Answer:

- Invulnerable
- Anti-authority
- Impulsive
- Macho
- Complacent
- Resigned
- Group Think
- Mission Focused
- Bias

EXERCISES

1. Identify a time in your career or in college that you were overwhelmed with stress. How did the stress impact you? What did you do to deal with the stress? Can these techniques be used on the fireline? If not, have you explored ways to deal with stress in a productive manner?

   Answer: Answer should be analytical in nature. Did the student address techniques to deal with stress? Did they provide a personal observation?

2. Develop a personal safety culture. If human error is at the root of most fatalities or injuries, then how do you propose to personally address your ability to make the proper and correct decisions to eliminate human error?

   Answer: Answer should be analytical and provide personal insight. The intent here is to stimulate and provoke critical analysis of human error as it relates to firefighting activities.

3. Have you personally witnessed a hazardous attitude in your career or in school? What was the attitude? Was the issue addressed and if so, how was it addressed? Was it effective?

   Answer: Answer should be more than yes or no. The student should provide an analysis and personal observation that will help them develop an approach to deal with hazardous attitudes.

CHAPTER 12

Knowledge Assessment

1. Identify the three categories of wildland fire hand tools.

   Answer:

   1. Cutting
   2. Scraping
   3. Smothering

2. True/False: Wildland firefighters should not inspect their hand tools to ensure they are effective and safe.

   Answer: False

3. When working on tools, the wildland firefighter should wear ________________, ________________, and ________________.

   Answer: Gloves, safety glasses, and long sleeved shirts.

4. Firefighters should maintain _____ feet between each firefighter while working and walking with hand tools.

   Answer: 10 feet

5. What is the most important consideration when walking with tools?

   Answer: Cover the cutting edge to reduce the potential for cuts to the firefighter.

6. What are the two approved methods to carry a tool while you are walking?

   Answer:

   - Hand tool by your side
   - Hand tool across your chest in the power zone

7. A firefighter should hold the tool and work with the tool in the ________________.

   Answer: power zone.

8. True/False: A sharp tool improves the effectiveness of a hand tool and reduces the wear and tear on the firefighter.

   Answer: True

9. Why should every firefighter carry at least one fusee?

   Answer: Firefighters need to carry a fusee in case they have to burn fuels out in the preparation of a safe area, deployment site, and/or emergency safety zone.

10. What is the fuel mixture for a drip torch?

    Answer: Four (4) parts diesel to one (1) part gas

CHAPTER 13

Knowledge Assessment

1. How many gallons of water does a backpack pump hold? How much extra weight is this to the firefighter?

Answer:

- 5 gallons
- +/-45 pounds

2. What are the four safety considerations when working with a backpack pump?

Answer:

1. Have other firefighters help you place the backpack onto your back.
2. Make sure all the caps are tightly secured.
3. Remove slack in backpack straps.
4. Avoid steep terrain and stepping over large obstacles.

3. Individual hose lengths are referred to as ________________.

Answer: Sections or sticks.

4. The three standard thread type for wildland fire hose ________________, ________________, and ________________.

Answer: NH: National Hose; NPSH: National Pipe Straight Hose; GHT: Garden Hose.

5. What are the three techniques used to unroll fire hose?

Answer:

1. Standard
2. Underhand/bowling
3. Overhand throw

6. What are the three techniques used to roll fire hose?

Answer:

1. Straight
2. Watermelon or melon
3. Figure 8

7. What direction should the male end of a hose be facing when using a standard hose rolling technique?

Answer: The male end should always be inside or to the interior of a hose lay.

8. What are the three main pumping sources used on wildland fires?

Answer:

1. Pump on an engine
2. Portable pump
3. Floater pump

9. Identify the two hose lays used on wildland fires.

Answer:

- Simple
- Progressive

10. What are the three nozzle patterns used during suppression activities?

Answer:

- Straight stream
- Fog or mist
- Combination or spray

EXERCISES

1. Practice the three hose rolling techniques identified in this chapter.

Answer: Student should be able to demonstrate all three hose rolling techniques.

2. Practice the three hose unrolling techniques identified in this chapter.

Answer: Student should be able to demonstrate all three hose unrolling techniques.

CHAPTER 14

Knowledge Assessment

1. What is the green side of a fireline and what is the black side of a fireline?

   Answer: The green side of a fire is considered the area outside of the control/fireline or the area that firefighters do not want to burn.

   Black side area inside the fireline/control line.

2. The basic principle for constructing fireline is: Construct fireline _____ times the sustained flame lengths.

   Answer: 1½

3. Fireline should be dug or constructed to ________________ soil.

   Answer: mineral

4. What are the five types of fireline used to contain a wildland fire?

   Answer:

   1. Hand line
   2. Machine line
   3. Wet line
   4. Retardant line
   5. Explosives line
   6. Black line and natural control lines (although this could be argued that it is not a true fireline, but a control line).

5. Described the one lick hand line construction.

   Answer: This technique involves each firefighter improving the efforts of the firefighter in front of him or her.

6. Describe the leap-frog hand line construction.

   Answer: In this concept, each firefighter is responsible for a specific piece of the fireline. The firefighter pioneers the line in his or her area and improves the line to the desired width. Everything from the first to last effort in a specific area is done by an individual firefighter.

7. What are the three primary attack methods used on a wildland fire?

   Answer:

   1. Direct
   2. Indirect
   3. Flanking/parallel

8. Firefighters cannot direct attack a fire if the flame lengths are over _____ ft. tall.

   Answer: Four

9. What is the fire suppression concept in which fuels that remain between the main fire and the control line are burned out? This action provides for safety of firefighters.

   Answer: Blackline concept

10. What type of attack method is Attack from the Black?

    Answer: Direct

11. What is the method of attack where fireline is constructed right on the fire perimeter? ("Keeping one foot in the black.")

    Answer: Direct

12. Define the term "burning out."

    Answer: Burning out is intentionally igniting fuels inside the fireline to eliminate unburned fuels.

13. Define the term "blackline concept."

    Answer: Blackline concept ensures firefighters have a solid edge of black directly against the fire or control line.

14. Define the term "patrolling."

    Answer: To go back and forth vigilantly over a length of control line during and/or after construction to prevent breakovers, suppress spot fires, and extinguish overlooked hot spots.

EXERCISES

1. On a piece of paper, illustrate the one lick hand line construction technique and list the pros and cons of this technique.

   Answer: Student should be able to draw this technique and list the pros and cons. Instructor discretion on pros and cons. Some examples include:

   Pros: Very fast

   Cons: Crews have a tendency to build the fireline too wide using this technique. Everybody wants to contribute to the team effort, so he or she can improve what the firefighter in front has done. When this is done, the end result may be a six-to-eight-foot wide fireline when a three-foot wide fireline was all that was needed. This ultimately creates an issue with repairing (rehabbing) the fireline.

2. On a piece of paper, illustrate the leap-frog line construction technique and list the pros and cons of this technique.

   Answer: Student should be able to draw this technique and list the pros and cons. Instructor discretion on pros and cons. Some examples include: Pros: effective in rocky terrain and in areas with high density of trees. Cons: not as fast as progressive or one-lick and individual firefighters may be tasked with more work than other crew members.

3. On a piece of paper, illustrate a patrol system used to detect spot fires.

   Answer: Students should be able to draw a patrol system. Instructor discretion.

CHAPTER 15

Knowledge Assessment

1. Wildland firefighters use what type of hand-held radio? What is the most common manufacturer of wildland fire radios?

   Answer:

   - Field-programmable, two-way radios
   - Bendix King

2. What are frequencies?

   Answer: Similar radio waves.

3. Are 10 Codes allowed on a wildland fire?

   Answer: No.

4. Why is mopup important to fire suppression?

   Answer: Mopup is an essential component of fire suppression and is often considered one of the last activities performed on a fire (with rehabilitation efforts being the other) before a fire is considered controlled or fully extinguished.

5. What are the two types of mopup performed on a wildland fire? Explain both.

   Answer:

   - Wet Mopup: Involves water and/or foam
   - Dry Mopup: Mopup without water and/or foam

6. What is the primary mopup system used on wildland fires?

   Answer: Gridding.

7. True or False: A wildland firefighter should not expect to encounter hazardous materials on a wildland fire.

   Answer: False.

8. What is the bottom line on dealing with hazardous materials on a wildland fire?

   Answer: Do not engage hazardous materials if you are not certified, trained, or equipped.

9. What is the most useful reference tool when dealing with hazardous materials?

   Answer: *Emergency Response Guidebook (ERG).*

10. What are illicit labs? Why are they harmful to firefighters?

    Answer: Drugs, chemical agents, explosives, and/or biological labs. They may have extensive booby traps or hazardous materials that can severely harm or kill firefighters.

11. What are the two categories of ignition types?

Answer:

- Natural
- Human-caused

12. What are the three primary duties related to fire investigations for initial attack resources?

Answer:

1. Observe suspicious activities
2. Document such activities
3. Protect the point of origin

13. Define a cultural resource.

Answer: Physical evidence or place of past human activity. A site, object, landscape, structure or a site, structure, landscape, object, or natural feature of significance to a group of people traditionally associated with it.

14. List the measures you can use to ensure cultural resources are maintained.

Answer:

- Do not disturb the site.
- Take GPS points and mark the location on a map.
- Flagging can be used to identify the sites.
- Advise your supervisor who should then advise fire managers.
- Avoid any tactical activities through or within the site.
- When in doubt, allow an expert (cultural resource specialist) to make the mitigation decision.

CHAPTER 16

Knowledge Assessment

1. Define the wildland-urban interface.

   Answer: The line, area, or zone where structures and other human development meet or intermingle with undeveloped wildland or vegetative fuels

2. True/False: A firefighter should do everything possible to protect structures at all costs.

   Answer: False: Firefighters should not risk their lives for a structure.

3. Identify five unique hazardous materials often found in the interface.

   Answer: Choose five from the following:

   - Propane tanks
   - Flammable liquids
   - Paint cans
   - Pesticides
   - Herbicides
   - Cleaning materials
   - Fuel tanks
   - Drums
   - Ammunition
   - Explosives
   - Vehicles
   - Plastics or synthetic material

4. What labs are considered illicit?

   Answer:

   - Drugs
   - Moonshine facilities
   - Chemical agents
   - Explosives
   - Biological labs

5. Identify the three tactical plans used in interface fires.

   Answer:

   1. Defensive
   2. Offensive
   3. Combined

6. What are four potential water sources used in interface fires?

Answer: Choose four from the following:

- Swimming pools
- Hot tubs
- Storage tanks
- Fish ponds
- Irrigation ditches,
- Hydrants
- Dry hydrants
- Standpipes

7. Define "structure triage."

Answer: Triage allows firefighters to establish priorities based on what structures can be protected from advancing wildfires.

8. What is the ultimate determining factor for triage?

Answer: Firefighter exposure and safety are the ultimate determining factors.

9. Identify the four categories of structural triage.

Answer:

- Defensible—stand alone
- Defensible—prep and hold
- Non-defensible—prep and leave
- Non-defensible—rescue drive by

10. True/False: Firefighters should not be considerate to home owners property and belongings?

Answer: False.

EXERCISES

1. Identify and explain the safety concerns with downed power lines.

Answer: The response should cover most of the following points:

- **Communicate:** Notify all responders of down electrical lines. Obtain radio check-back.
- **Identify:** Determine entire extent of hazard by visually tracking all lines, two poles in each direction, from the downed wire.
- **Isolate:** Flag area around down wire hazards; post guards.
- **Deny entry:** Delay firefighting actions until hazard identification and flagging is complete and/or confine actions to safe areas.
- **Downed line on vehicle:** Stay in vehicle until the power company arrives. If vehicle is on fire, jump out with both feet together. Do not touch the vehicle. Keep feet together and shuffle or hop away.
- **Always treat downed wires as energized!**

2. Research the term "defensible space." What does the term mean? How does it support triage decisions. List your reference(s).

Answer: Students should provide a definition or explain the concept of defensible space around a structure.

Students should explain that defensible space removes flammable material next to or around structures which helps determine if the home can be protected.

References should be provided and evaluated by the instructor.

3. Research Firewise Communities. What are they? How do Firewise Communities help firefighters protect structures? List your reference(s).

Answer:

- Students should explain Firewise Communities.
- Instructor discretion on second part.
- References should be provided and evaluated by the instructor.

CHAPTER 17

Knowledge Assessment

1. Identify five types of maps used on wildland fires.

   Answer: Select from the following:

   - Topographic
   - Subdivision
   - Road (highway)
   - Incident action plan
   - Facilities
   - Infrared
   - Land ownership
   - Operational briefing
   - Public information
   - Situation
   - Fire progression
   - Orthoimagery

2. Define a contour line.

   Answer: Actual lines on a map along which every point is at the same height above sea level.

3. What do contour lines that make V's indicate?

   Answer: Water feature. V will point upstream.

4. What do contour lines that make U's indicate?

   Answer: Identifies a ridge. U will point down the ridge.

5. What land descriptive system is most used by fire managers?

   Answer: Public Land Survey System

6. The PLSS is based on three concepts. What are they?

   Answer:

   - Township
   - Range
   - Section(s)

7. A township is how many square miles?

   Answer: 36

8. A section is how many square miles and how many acres?

   Answer: 1 square mile—640 acres

9. An acre is about the size of a ________________. An acre consists of __________ square feet.

   Answer:

   - Football field.
   - 43,560

10. True/False: A compass provides a north reading in relation to true north.

    Answer: False. It provides a reading to magnetic north.

11. GPS units identify a location based on ________________.

    Answer: Latitude and longitude.

EXERCISES

1. Complete the following exercise using the PLSS system of legal land description. A mom and dad own one (1) section of land. They have four sons. They want to give each son the equal amount of property; however, they want to keep a quarter section for themselves. Illustrate how you would equally divide the property for each son and the mom and dad. Do not provide a math solution. Draw the solution.

Mom/Dad	Mom/Dad	Brother #1	Brother #1
Mom/Dad	Mom/Dad	Brother #1	Brother #2
Brother #4	Brother #4	Brother #2	Brother #2
Brother #4	Brother #3	Brother #3	Brother #3

Mom/Dad will keep ¼ section or 160 acres.
Each brother (son) receives three-quarter sections or 120 acres.

INDEX

A

B

C

D

E

F

G

H

I

L

N

O

P

R

S

T

U

V

W

CPSIA information can be obtained
at www.ICGtesting.com
Printed in the USA
JSHW042012210722
28384JS00003B/12